T0140185

Studies in Computational Intelligence

Volume 662

Series editor

Janusz Kacprzyk, Polish Academy of Sciences, Warsaw, Poland
e-mail: kacprzyk@ibspan.waw.pl

About this Series

The series "Studies in Computational Intelligence" (SCI) publishes new developments and advances in the various areas of computational intelligence—quickly and with a high quality. The intent is to cover the theory, applications, and design methods of computational intelligence, as embedded in the fields of engineering, computer science, physics and life sciences, as well as the methodologies behind them. The series contains monographs, lecture notes and edited volumes in computational intelligence spanning the areas of neural networks, connectionist systems, genetic algorithms, evolutionary computation, artificial intelligence, cellular automata, self-organizing systems, soft computing, fuzzy systems, and hybrid intelligent systems. Of particular value to both the contributors and the readership are the short publication timeframe and the worldwide distribution, which enable both wide and rapid dissemination of research output.

More information about this series at http://www.springer.com/series/7092

Michael Emmerich · André Deutz
Oliver Schütze · Pierrick Legrand
Emilia Tantar · Alexandru-Adrian Tantar
Editors

EVOLVE – A Bridge between Probability, Set Oriented Numerics and Evolutionary Computation VII

 Springer

Editors

Michael Emmerich
Leiden Institute of Advanced Computer Science
Leiden University
Leiden
The Netherlands

André Deutz
Leiden Institute of Advanced Computer Science
Leiden University
Leiden
The Netherlands

Oliver Schütze
Departamento de Ingeniería Eléctrica
CINVESTAV-IPN
Mexico City, D.F.
Mexico

Pierrick Legrand
Institut de Mathématiques de Bordeaux
University of Bordeaux
Bordeaux Cedex
France

Emilia Tantar
Luxembourg Centre for Systems Biomedicine
University of Luxembourg
Belval
Luxembourg

and

Interdisciplinary Centre for Security, Reliability
and Trust
University of Luxembourg
Luxembourg
Luxembourg

Alexandru-Adrian Tantar
Computer Science and Communications Research
Unit
University of Luxembourg
Luxembourg
Luxembourg

and

Interdisciplinary Centre for Security, Reliability
and Trust
University of Luxembourg
Luxembourg
Luxembourg

ISSN 1860-949X   ISSN 1860-9503   (electronic)
Studies in Computational Intelligence
ISBN 978-3-319-84133-5   ISBN 978-3-319-49325-1   (eBook)
DOI 10.1007/978-3-319-49325-1

Printed on acid-free paper

This Springer imprint is published by Springer Nature
The registered company is Springer International Publishing AG
The registered company address is: Gewerbestrasse 11, 6330 Cham, Switzerland

Preface

Numerical and computational methods for solving (multiobjective) optimization, game theory, and machine learning problems are actively researched in recent years. During the last decades, various schools of deterministic and stochastic algorithm research have emerged. In order to solve problems in practice reliably and efficiently, there is a need for work across methodological boundaries.

This book comprises nine selected works on this topic. The work is by participants of the EVOLVE 2013 conference held in July 2013 at Leiden University, The Netherlands, from various fields of science such as computer science, mathematics, and engineering. This book's chapters are peer-reviewed by an international review panel. They provide extended versions of selected papers from the contributions to the conference.

This resulting book includes original work by the authors and covers important topics in both theory and applications, for instance, the role of diversity in optimization, statistical approaches to combinatorial optimization, computational game theory, and cell mapping techniques for numerical landscape exploration. Applications focus on aspects such as robustness, handling multiple objectives, and complex search spaces in engineering design and computational biology.

We wish our readers interesting insights from this book and inspirations for their own research and work on problem solving.

On behalf of the editors
Michael Emmerich and André Deutz
Conference Logo

Contents

Part I
Set Oriented Optimization Methods

A Survey of Diversity Oriented Optimization: Problems, Indicators, and Algorithms

Vitor Basto-Fernandes, Iryna Yevseyeva, André Deutz
and Michael Emmerich

Abstract In this chapter it is discussed, how the concept of diversity plays a crucial role in contemporary (multi-objective) optimization algorithms. It is shown that diversity maintenance can have a different purpose, such as improving global convergence reliability or finding alternative solutions to a (multi-objective) optimization problem. Moreover, different algorithms are reviewed that put special emphasis on diversity maintenance, such as multicriteria evolutionary optimization algorithms, multimodal optimization, artificial immune systems, and techniques from set oriented numerics. Diversity maintenance enters in different search operators and is used for different reasons in these algorithms. Among them we highlight evolutionary, swarm-based, artificial immune system-based, and indicator-based approaches to diversity optimization. In order to understand indicator-based approaches, we will review some of the most common diversity indices that can be used to quantitatively assess diversity. Based on the discussion, 'diversity oriented optimization' is suggested as a term encompassing optimization techniques that adress diversity maintainance as a major ingredient of the search paradigm. To bring order into all these different approaches, an ontology on diversity oriented optimization is proposed. It provides a systematic overview of the various concepts, methods, and applications and it can be extended in future work.

V. Basto-Fernandes (✉)
Polytechnic Institute of Leiria, Instituto Universitário De Lisboa (ISCTE-IUL), University
Institute of Lisbon (ISTAR-IUL), 1649-026 Lisboa, Portugal
e-mail: Vitor.Basto.Fernandes@iscte.pt

I. Yevseyeva
Faculty of Technology, School of Computer Science and Informatics, De Montfort University,
Leicester LE1 9BH, UK
e-mail: iryna.yevseyeva@dmu.ac.uk

A. Deutz · M. Emmerich
Leiden University, 2333 CA Leiden, The Netherlands
e-mail: a.h.deutz@liacs.leidenuniv.nl

M. Emmerich
e-mail: m.t.m.emmerich@liacs.leidenuniv.nl

© Springer International Publishing AG 2017
M. Emmerich et al. (eds.), *EVOLVE – A Bridge Between Probability,*
Set Oriented Numerics and Evolutionary Computation VII,
Studies in Computational Intelligence 662, DOI 10.1007/978-3-319-49325-1_1

1 Introduction and Motivation

The concept of *diversity* plays a crucial role in various optimization and search techniques. Diversity maintenance can help to find a globally optimal solution, but it might also be the goal of optimization to produce a diversified set. Strategies to maintain diversity are used in various methods, in particular in population-based metaheuristics and their variation and selection operators. Moreover, there exists a multitude of diversity measures, addressing different aspects of what common sense might tell us what diversity is.

In this chapter, we look at the concept of diversity across several different methods and try to define 'diversity oriented optimization' as an emerging topic in optimization methods. Towards this end, we propose an ontology that seeks to provide a systematic overview, and can be used by the algorithm community to identify essential similarities and differences between different methods. This can be useful to find related work across algorithmic sub-disciplines or to identify prevalent trends in the field.

Before giving more concrete definitions of diversity, a tentative definition of diversity could be given as follows: Diversity is a property of a multi-set the elements of which are all members of the same space, say $\mathbb{M}$. The space can be, for instance, the set of integer numbers, the set of real vectors of dimensions n, or, the set of all molecular structures. It is demanded that $\mathbb{M}$ is at least equipped with a dissimilarity measure $d : \mathbb{M} \times \mathbb{M} \to \mathbb{R}_0^+$. Intuitively, we would then say that a subset of $\mathbb{M}$ is more diverse than another subset of $\mathbb{M}$, if

1. it contains more different elements,
2. elements are more different with respect to each other,
3. and more evenly distributed over $\mathbb{M}$.

In the literature a diverse set of diversity measures has been suggested, emphasizing these or subsets of these three aspects.

Traditionally, formal definitions of diversity have been mainly investigated in biological statistics in order to measure population diversity, but recently there is a growing interest in other fields of science and economics, too. Examples are cultural sciences, innovation management, and financial portfolio theory. Last but not least, the concept of diversity is a concept of vital importance in contemporary optimization algorithms. In the following we will focus on this last mentioned topic. Thereby we will often refer to terminology and diversity measures developed in other fields of science.

In many optimization techniques it is not even made explicit which diversity measure is used. Rather it is claimed, that a certain operator or strategy is used to increase diversity or to maintain diversity, not being explicit what exactly is meant by diversity. This is however not the case in the so-called indicator-based optimization methods, which aim for improving a diversity measure that is defined *a priori*. In this chapter we will therefore first review methods that use a rather vague definition of diversity, and then introduce indicator-based methods that refer to exact definitions of diversity and review these definitions.

This work is structured in three parts:

- The first part is focused on different optimization problems and how the concept of diversity is important to solve or define these problems.
- The second part reviews optimization methods that emphasize the concept of diversity for various reasons. The section on indicator-based methods also discusses various measures of diversity.
- Finally, in the third part, an ontology that integrates the different theoretical concepts, methods, and applications is developed. Based on this ontology, commonalities and essential differences between methods and problem definitions are discussed.

Figure 1 presents an overall perspective of all dimensions of diversity oriented optimization addressed in this study.

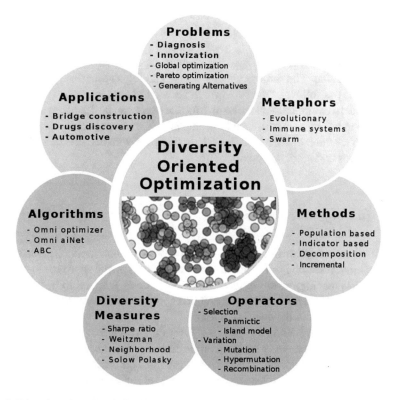

Fig. 1 Diversity oriented optimization. The lists serve as examples but are not exhaustive

2   Problem Domains in Diversity Oriented Optimization

This section reviews several problem domains, in which diversity is of importance to increase the performance of optimization methods or in which diversity is aimed for in the results of optimization methods.

2.1   Generating Alternatives

In engineering problems, when an algorithm provides the decision maker with a solution of a *single objective* optimization problem, it might not always correspond to the decision maker's preferences. In order to satisfy a demanding decision maker, or several decision makers, the developers of Evolutionary Algorithm to Generate Alternatives (EAGAs) [47] suggested an algorithm that searches for several good (not necessarily optimal) but maximally different solutions.

In complex problem fields and search spaces, such as drug discovery, only by looking at a solution it can be judged by a domain expert, whether or not that solution is possibly suitable. The famous chemist Linus Pauling summarized the process of discovery [25] as follows: "*the best way to get a good idea is to get a lot of ideas*". Indeed, modern computational tools for drug discovery can be described as diversity-oriented search for generating a set of promising alternatives [42]. A similar view is taken in the research of Ulrich et al. [38], on finding diversified sets of architectural bridge designs and hardware configurations.

2.2   Multiobjective Optimization

In multi-objective optimization it is a common practice to compute a diverse set of Pareto optimal solutions. As opposed to the so-called *a priori* approach to multicriteria decision making, where different objectives are aggregated to a scalar utility value, the so-called *a posteriori* approach first computes all Pareto optimal solutions and presents this set to the decision maker.

As pointed out by Knowles [21], the Pareto optimal set can be viewed as the set of optimal solutions over all meaningful linear or non-linear scalarizing utility functions. The knowledge of the Pareto front provides the decision maker with information on the trade-offs between different objectives and the availability of solutions satisfying potential goal vectors. Typically, diversity is measured for the set of non-dominated objective function vectors, but more recently the importance of decision space diversity has been stressed in several publications [8, 29, 31, 40]. Here the idea is, that it can be important to compute different pre-images of the same preferred point on the Pareto front, given they exist. For instance, a decision maker might prefer some solutions to others for subjective, non-explicit reasons.

2.3 Innovization

By analyzing diversified sets of good (not necessary optimal) solutions, provided by a diversity-oriented search algorithm, a designer may learn some important properties of the decision space. The findings might even be generalizable and lead to the discovery of some design patterns. It could be that some properties are shared among high-performing solutions. Early research in this direction was conducted by Ian Parmee et al. [24] in the field of evolutionary design optimization using so-called cluster oriented genetic algorithms (COGAs).

Recently, the derivation of design principles from optimization results has been discussed in the context of multi-objective optimization. The question is here how a design changes when navigating along the Pareto front and whether solutions on the Pareto front are distinguished from dominated solutions. Modern data mining and data analysis techniques are used to discover patterns and rules. Deb et al. call this process of innovation by optimization '*innovization*', see e.g. [10–12] and it is believed that this methodology has a large potential to revolutionize the engineering design process in industry.

2.4 Novelty and Interestingness

In this context it is also important to consider resistance of decision makers to make large changes to existing solutions. This resistance is explained by the risk due to the lack of knowledge about a novel design. In [27], the concepts such as *interestingness* and *novelty* are defined in the context of innovization. It is emphasized that candidate solutions for new designs must be different to known solutions but still be understood well enough on the basis of existing models used in the domain.

In the interestingness measure derived by Reehuis et al. [27], diversity is not only seen as an asset but it also comes with a risk, due to the unknown properties of novel designs. It is stressed that a certain balance between novelty and sticking to known and well-tested designs must be maintained, in order to make solutions appear interesting to designers. Moreover, interestingness can only be measured relative to existing designs.

2.5 Finding Peaks in Multimodal Landscapes

When exploring *multimodal* landscapes for the global optimum, it might be beneficial to explore several attractor basins, or peaks, in parallel. Many optimization strategies focus, after a transient initial phase, only on a single attractor. This also holds for population-based algorithms due to several causes of diversity-loss [28]. Diversity maintenance techniques can be crucial to improve the probability of finding the global optimum. Among others, Stoean et al. [36] argue about the need of active diversity maintenance to be considered in the context of global optimization.

In [12] the need of finding *all* global optima (multi-global optimization) is termed *multi-global optimization*, both in single- and multi-objective optimization. Moreover, algorithms that seek to gain knowledge about the topological structure of multimodal landscapes are recently developed [26]. The goal of these methods is not only to discover global optimizers but also to find local optimizers and their connectedness.

2.6 Model-Based Diagnostics

A system of cause and effect elements can be modeled as a function from input variables (causes) to output variables (effects). When performing model-based fault diagnosis, it is important to find all possible causes to an observed effect. If there are different possible causes, this indicates that further data is needed to identify the true cause. This aspect has been discussed in [48]. Here a diversity-oriented evolutionary algorithm was developed for finding potential contaminant sources in water distribution networks.

2.7 Dynamic and Robust Optimization

Both, when the goal is to find robust optima or when the positions of optima change over time, the maintenance of diversity can be a crucial component of the algorithm. For instance, in [19], it was pointed out that the lack of diversity can lead to stagnation of population-based search in suboptimal regions in dynamically changing environments. The idea that diversity makes a group of individuals more robust to attacks by, for instance, viruses is a common explanation for the evolutionary benefits of diversity maintenance. When transferred to optimization, such considerations might play a role when it comes to selection of portfolios or teams, which will be discussed later in Sect. 5.

2.8 Summary

In this section various motivations for maintaining diversity in optimization were reviewed. Across these different motivations, it is widely common that diversity is seen as an asset. It is either useful to maintain diversity in order to find a better result, or the result itself should include different solutions in order to be more interesting for the decision maker.

Negative aspects of diversity are that a too large output set might confuse the decision maker or that solutions that are too different from existing designs might not be trusted. In both cases it would be a rather good strategy to use diversity maintenance in moderation, rather than abandoning it at all.

3 Bio-Inspired Methods for Diversity Oriented Optimization

Next, it will be discussed how the problem of diversity maintenance is adressed in optimization algorithms.

Often, the mechanism has been inspired by nature and therefore the first part of the discussion of the methods will be devoted to nature inspired methods. In nature inspired methods terminology is often related to metaphors from biology. This, on the one hand makes it easy to see the analogy, on the other hand it can make it difficult to compare different bio-inspired methods with each other. Besides reviewing mechanisms, this section will also reveal commonalities between various strategies which might be slightly obscured by terminology.

3.1 Evolutionary Algorithms

In *Evolutionary Algorithms (EAs)*, a population (set) of individuals (solution candidates) evolves over time to a population with a better average objective function value among individuals. The process is driven by selection of promising parents, recombination and mutation.

Using a population of individuals by itself does not guarantee however that diversity is maintained. Even in situations where individuals share the same objective function value, that is optimization on a plateau function, diversity is quickly lost due to genetic drift [28]. Those individuals of types that are underrepresented in the populations tend to reproduce less frequently and the number of individuals of this type gets even smaller in subsequent generations. This effect can be quantified by Markov chain models (cf. [28]) and the time to extinction of an individuals tends to be proportional to the population size.

A common paradigm for maintaining diversity in EA populations is that of niching. For instance, *niching* [32] allows selecting only few solutions located within the same region of the objective or decision space, for parental or environmental selection. As an example of a global optimization strategy with diversity maintenance the omni-optimizer [13] will be discussed in more detail.

3.1.1 Omni-Optimizer

The omni-optimizer [13] was developed to allow both single and multi-objective optimization by means of a single generic evolutionary algorithm. It is based on the well-known generational NSGA-II algorithm for finding all Pareto optimal solutions for problems with multiple, usually conflicting objectives. However, omni-optimizer can also adapt automatically when simpler cases of multi-objective problems with a single optimum are detected, or when a single objective optimization problem should be solved either for finding multiple optima or a single optimum.

When compared to the original NSGA-II, omni-optimizer is improved with restricted mating selection of similar individuals competing in tournament. For tournament, two pairs of individuals are selected such that for each randomly selected individual its nearest neighbour in objective space becomes its competitor. Such restricted parent selection in tournaments limits competition to very similar individuals only, it also preserves possible multiple optima and speeds up convergence to them or to a single global optimum. A modified environmental selection is computed not only taking into account the phenotype (objective space evaluation as in NSGA-II), but also the genotype (decision space) of an individual. Moreover, pairwise comparison of individuals is based on a modified ε-domination evaluation, which neglects a small differences on objective function values between two individuals when deciding which one is the best. In addition, a more disruptive mutation operator is obtained by modifying the treatment of the original polynomial mutation at the boundaries of variables. The algorithm's initial population is based on Latin-hypercube random uniform sampling, but predefined sampling can also be used.

For the multi-objective case the winner of tournament is selected taking into account feasibility, constraints violation, dominance and crowding of each individual evaluated in objective space only. When selecting among two individuals the following criteria apply: (1) A feasible individual is preferred to an infeasible one; (2) A feasible non-dominated individual is preferred to a feasible dominated one; (3) Among two feasible non-dominated individuals the one with higher crowding distance is preferred (or randomly selected if crowding distance is the same); And (4) among two infeasible individuals the one with smallest constraint violation is preferred. When compared to NSGA-II, omni-optimizer evaluates dominance relation using ε-dominance. The advantage of this type of domination is its tolerance towards small differences of near non-dominated individuals. Keeping such individuals may be beneficial when diversity is an issue and when several rather than one solutions should be selected. A penalty for constraints violation is computed as a sum of violations of all equality and inequality constraints. In case of single objective optimization, the tournament selection degenerates to the above mentioned criteria (1) and (4), and feasible individuals with smallest objective function values are preferred.

For individuals represented by real-coded variables, SBX crossover (on half of variables, on average) and a modified polynomial mutation operators are used. When compared to the original polynomial mutation operator, which has the disadvantage of having no effect as soon as a variable reaches its boundary, the new polynomial mutation operator assigns non-zero probability of mutation even if one of the variables of an individual is on its boundary. For individuals represented by binary coded variables two-point crossover is applied together with bitwise mutation.

Although omni-optimizer follows a $(\mu + \lambda)$ schema, similar to NSGA-II, the environmental selection stage is modified. Omni-optimizer considers similar principles used for parents selection, such as feasibility, constraints violation, domination and crowding of individuals in objective space, but applies ε-domination similar to that in mating selection. Moreover, crowding distance is computed as an average between its two closest neighbours for both objective and decision spaces on all decision variables and objectives values, respectively. The biggest of the two values, either its normalized crowding distance in objective space or its normalized crowding distance in decision space, is taken as a crowding distance of an individual. By considering decision space, this crowding distance allows differentiating between two non-dominated individuals with the same/similar evaluation in objective space, but different structurally (in decision space).

The results of omni-optimizer testing on a number of single uni-optimal and multi-optimal test problems and multi-objective uni-optimal and multi-optimal test functions are reported in [13]. They are promising both in terms of quality and coverage of the Pareto front in multi-objective optimization problems.

However, omni-optimizer reveals poor results of crowding distance in three and more dimensions and a possible loss of optimal solutions when using ε-domination. Moreover, the evident advantage of omni-optimizer developed as a generalized solver to deal with a variety of problems does not exclude the fact that specific problem-oriented algorithms may be more efficient for specific problems. For instance, omni-optimizer was not designed to find local optima of multi-modal problems but only to find global one(s).

3.2   Artificial Immune Systems

Immune Systems have inspired various algorithms, among which are also algorithms in diversity oriented search. These so called Artificial Immune Systems (AIS) have been proposed by de Castro and Von Zuben [5]. In AISs, adaptation happens by cumulative variation and selection within cells (cloning and hypermutation): In the biological counterpart, immunoglobulin nucleotides are randomly inserted and deleted from recombined immunoglobulin gene segments. In AISs mutation operators (hypermutation and receptor editing) are applied to vectors that correspond to lymphocyte clones and might involve exchange or shifts of positions of individual representation data elements. These AIS operators are of major importance for cell diversification and affinity enhancement to antigens.

In AISs, diversity is preserved intrinsically by *clonal selection theory* [4] and *immune network theory* [18] principles. In AISs, a population evolves by cloning and mutation (hypermutation) processes, with genetic variability inversely proportional to affinity and concentration among individuals in the population. Self-adapting metrics, such as *affinity* among cells guarantee that concentration of similar cells decrease in the presence of better neighboring cells: the closer and higher the concentration of better cells, the higher their influence. Cells without better solutions in their neighborhood increase their concentration proportionally to their fitness.

3.2.1 Artificial Immune Systems *versus* Evolutionary approaches

Although coming from two different theoretical biology domains, AISs and evolutionary systems inspired by evolution theory, follow some similar fundamental principles: diversity, natural variation and selection are both present.

In both EAs and AISs, the population evolves by variation of selected cells/ individuals. Genetic crossover and mutation are responsible for population diversity and fine-tuning. These variation mechanisms are present in both AISs and evolutionary approaches. In particular, in EAs, biological evolution happens by cumulative natural selection among individuals: crossover and mutation generates offspring from mixing parental genes.

When compared to other nature-inspired algorithms, *Artificial Immune Systems* (AIS)s automatically adjust the population size at each iteration depending on the problem needs and still preserves diversity of the shrinked or enlarged population. Moreover, they apply hypermutation, which can be seen as a concept that combines aspects of mutation and recombination.

Both approaches are successfully applied for diversity-oriented search, e.g. in multi-objective optimization, see e.g., [8, 31].

3.2.2 Omni-aiNet

Omni-aiNet [7] was developed to serve the same purpose as omni-optimizer: for solving single and multi-objective optimization. However, contrary to omni-optimizer, omni-aiNet is based not on evolutionary algorithms, but on artificial immune systems.

Omni-aiNet is based on opt-aiNet [9] and several AISs principles, such as *clonal selection principle* and *immune network theory*, and adapts some ways of solving a common optimization dilemma of driving search towards both exploration and exploitation of the objective space. The algorithm starts by initializing a randomly user-predefined number of individuals with a set of real-coded variables. The initial population enters the generational loop, at each iteration of which the current population goes through cloning, hypermutation, selection and gene duplication stages, until some stopping criterion is met. At the cloning stage for each individual of the current population user-predefined number of clones are created and mutated by polynomial mutation. The probability of mutation is selected inversely proportional to affinity of a clone with its antigen. This mutation parameter is defined empirically.

From the current population and the population of mutated clones the new population is selected similarly to the parental selection of omni-optimizer (based on ε-domination and constraints violation principles), except that instead of the crowding distance principle for selecting among two mutually non-dominating solutions, grid-based selection is used, similarly to the one introduced in the Pareto Archived Evolution Strategy (PAES) [22]. For each objective the interval between its minimal and maximal values is divided into a grid with user defined resolution. After partitioning the objective space into a grid of cells according to all objectives, only a user-predefined number of solutions closest to the centre of each cell is selected (here this number is equal to the number of grids).

3.3 Swarm Intelligence

Swarm Intelligence (SI) is another biological paradigm similar to AISs dynamic approach to diversity preservation: swarm self-adapts itself to the environment and/or communication between swarm members-agents when necessary [2]. For instance, in *Artificial Bee Colony* (ABC) algorithms, a special type of swarm bee-agents, called *scouts*, are activated on the last exploration stage of the algorithm to promote following diversification. After two exploitation-intensification phases, where the *employed* and *onlooker* bee-agents search for local optima, based on deterministic and probabilistic selection, respectively, *scout* bee-agents force abandoning of non-promising solutions and start exploring new solutions corresponding to new decision space regions. A similar multi-agent approach for multi-modal search, motivated by exploration strategies of *scouts*, is the *self-organizing scouts* (SOS) algorithm [3].

Biological paradigms are also addressed in spatial population structures as opposed to panmictic ones. Examples are cellular genetic algorithms [1] and spatial predator-prey algorithms in multi-objective optimization, which were investigated first in [23].

Besides biological paradigms, the mathematical programming community has developed several algorithms for diversity-oriented optimization that exploit mathematical structures of functions expressing diversity [46].

In Table 1 we present a summarized comparison of the bio-inspired metaphors that were described in previous sections.

Table 1 Bio-Inspired Computational Metaphors

	Metaphors		
	EA	AIS	SI
Rationale	*Natural selection*	*Clonal selection*	*Social cooperation*
Reproduction	*Recombination*	*Cloning*	*Specialization*
Local search	*Mutation*	*Hypermutation*	*Onlooker individuals*
Variation	*Recombination*	*Cumulative variation*	*Scout individuals*
Adaptation	*Natural selection*	*Immunology principles*	*Multi-agent*
Diversity	*Niching, Various*	*Variable pop. size*	*Locality*

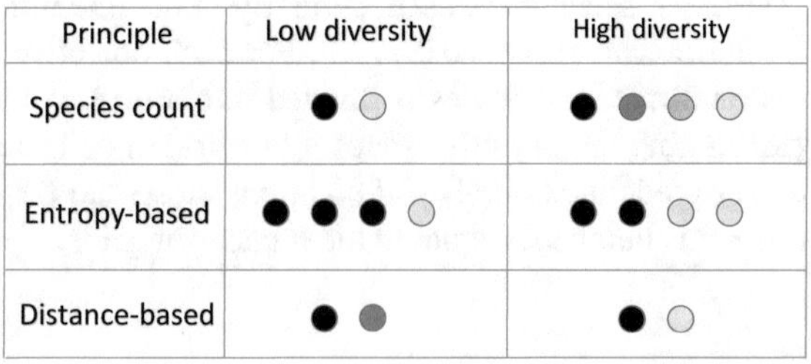

Fig. 2 Principles used in the definition of diversity. Individuals with different *gray* level are considered to be of different species. The more different the *gray* level, the more distant are the species to each other

Principle	Low diversity	High diversity
Species count	● ○	● ● ○ ○
Entropy-based	● ● ● ○	● ● ○ ○
Distance-based	● ●	● ○

4 Diversity Measures

When assessing the performance of diversity maintenance mechanisms, it is of crucial importance to apply reliable and well-understood diversity measures. Many of these measures stem from biostatistics, because in conservational biology it is a common problem to measure the diversity of species in a population. In diversity oriented optimization, with *species* we express the concept of a class of points that is essentially different from all other points in a population while being similar to each other. An overview of the underlying principles of diversity measures is provided in Fig. 2, where points with different colors symbolize different species. Diversity indicators might take into account simply the number of (essentially different) species in a population, or also measure the eveness of the distribution of species (entropy-based), or the dissimilarity of species with respect to each other. Next, we will discuss important diversity measures and their relation to each other in more detail.

4.1 Diversity Based on the Abundance of Species

The simplest diversity measure is *species richness*, which is just the total number of species in a population. This index has the drawback that it does not measure the relative abundance of a species. In other words, if a population with n species would almost entirely consist of individuals of the same species, it would have the same richness as a population with evenly distributed abundance of species. A couple of diversity indices presented next seek to circumvent this problem:

A classical diversity measure that takes into account the relative abundance of a species is the *Simpson index* [33]. For a given population P, the Simpson index $S(P)$ measures the probability that if we draw a random sample of an individual of a certain species without replacement, in a second experiment we draw an individual of a different species. If the distribution of species abundances is more even, this index has a higher value. Let n_i denote the number of individuals of species i, N denote the total number of individuals, and n the number of species. Then

$$S(P) = 1 - \frac{\sum_{i=1}^{n} n_i(n_i - 1)}{N(N - 1)}.$$

A disadvantage of the Simpson index is, that it is difficult to interpret for large populations, because the probabilities tend to get very close to 0. It is difficult for humans to judge the difference between a probability of, e.g., 0.0099 and of, e.g., 0.00999.

A similar measure is the Shannon entropy:

$$S(P) = -\sum_{i=1}^{n} \frac{n_i}{N} \cdot \log \frac{n_i}{N}.$$

The Shannon entropy reaches its maximum, if the species are equally abundant and grows with the number of species. Still the growth is limited by the slow growth of the logarithm.

The *Rich Gini Simpson quadratic index* (RGS) [17] obtains values from 0 to $n - 1$. The maximum is obtained for an evenly distributed population. It, thus, gives a good idea about species diversity, and can be compared for populations of different sizes. It is computed as:

$$RGS(P) = n \sum_{i=1}^{n} \frac{n_i}{N}(1 - n_i/N) = n\left(1 - \sum_{i=1}^{n} \left(\frac{n_i}{N}\right)^2\right).$$

The evenness of a distribution of species abundances could also be measured independently of the population size by the *Gini index*, which originated from measuring welfare of an economy [6]. It is given by the average absolute deviation from the mean and tends to zero as the population tends to be evenly distributed over the species.

$$G(P) = \frac{1}{n - 1}\left(n + 1 - 2\left(\frac{\sum_{i=1}^{n}(n + 1 - i)n_i}{\sum_{i=1}^{n} n_i}\right)\right).$$

4.2 Diversity Based on Distances

In the aforementioned methods, the distance between species is not considered. For instance Weitzman [43] suggested to consider sets with the same number of species but bigger dissimilarity between species to be more diverse. In addition, he demanded that a diversity measure should grow if the number of species increases. Based on these and a number of additional properties, he suggested a diversity measure that we will term *Weitzman Diversity* $W(P)$. It is defined as follows: Given a population P and a distance matrix d_{ij} for members i and j, $i = 1, \ldots, N$, $j = 1, \ldots, N$: Let $d(i, Q)$ be the distance between i and the closest element (aka neighbor) in Q, for some non-empty $Q \subset P$. Moreover, let $P \setminus i$ define the population with the i-th individual removed. Then, the Weitzman diversity is recursively defined via

$$W(P) := \max_{i \in \{1,\dots,N\}} (W(P \setminus i) + d(i, P \setminus i)).$$

The Weitzman diversity has interesting theoretical properties, but it is costly to compute it in practice for large populations. The running time of the fastest known algorithm scales with $O(2^N)$.

A simplification of the Weitzman distance would be to compute only the first iteration of the recursion, which is known as the MAX-MIN diversity [15]:

$$M(P) := \max_{i \in \{1,\dots,N\}} (d(i, P \setminus i)).$$

The MAX-MIN diversity is used as a straightforward way to measure diversity of subsets in operations research. It was shown that the problem of finding a k-size subset from P of maximal MAX-MIN diversity is NP hard. The MAX-MIN diversity can be used to compare populations that have the same size. Its maximum often yields evenly distributed populations, because every point seeks to maximize the distance to its nearest neighbor. However, in relative comparisons it might be misleading. Consider for instance a population $P_1 = \{1, 2, 8, 9\}$ and a population $P_2 = \{1, 2, 3, 4\}$. Then $M(P_1) = M(P_2)$, if we consider the absolute deviation as a distance measure. However, clearly P_1 is more widespread than P_2.

A proposal for a distance measure that shares most properties of the Weitzman diversity, but can be computed faster, is the *Solow Polasky diversity* [35]. It requires, however, a parameter $\theta > 0$ that needs to be chosen by the user. The definition of the Solow Polasky diversity $SP(P)$ is as follows: Let $c_{ij} = \exp(-\theta d_{ij})$ denote a correlation between point i and point j. If two points are of the same species the correlation is one. Let $\mathbf{M} = \mathbf{C}^{-1}$, assuming that $\mathbf{C}$ is of full rank. Then

$$SP(P) = \sum_{i=1}^{N} \sum_{j=1}^{N} m_{ij}.$$

It is easy to show that $SP(P)$ tends to N, if the distance between all species tends to be very large. Moreover, $SP(P)$ tends to one, if species are very similar with respect to each other. The parameter θ determines how fast the population tends to N when the distances increase.

In the literature also other distance-based diversity measures occur. For instance, the variance $V(P)$ of a population is given by $\mathbf{1}^T \Sigma \mathbf{1}$, where Σ is a problem specific covariance matrix. The entries of the covariance matrix can also depend on the dissimilarity. The variance of a population is often used in portfolio theory in order to measure the variance of the return (or risk) of a portfolio [44].

Another way to measure diversity is to compute arithmetic or geometric averages of distances to nearest neighbors. This measure can be computed efficiently and maximizing it will also lead to evenly spaced populations without duplicates. These so-called gap measures were discussed in [14] and used in various contexts in optimization. They, however, can only be used to compare populations of the same size.

It might be somewhat tempting to use the sum of pairwise distances as a diversity measure. However, when maximizing such measures, often clustering at the boundary is obtained. For instance, the population {1, 1, 4, 4} has a higher sum of pairwise distances than {1, 2, 3, 4}. The first population has a sum of pairwise distances of 24 while the second population has a sum of pairwise distances of only 20.

In the next section we will discuss so-called indicator-based algorithms that are directly oriented towards maximizing these diversity measures.

5 Indicator-Based Optimization Methods

A recent trend is to stress quality performance measures in the design of an algorithm. In the context of metaheuristics, methods that directly seek to maximize some quality indicator of a set are called indicator-based optimization algorithms [49]. Originating from methods that seek to find approximations to Pareto fronts, indicator-based optimization algorithms are recently also used in other domains.

Quality performance measures targeted to optimization algorithms need to cope with the fact that, the outcome is not constituted by a single solution, but by a set of solutions (e.g., in multimodal and multi-objective optimization problems).

Each of the indicators allows comparing algorithms with respect to one of several properties, among which are the quality of sets of individuals, diversity and distance to the optimal set (assuming it is known). At the same time, researchers noticed that optimizing indicators themselves is a good strategy for population evolution in the framework of an algorithm, and suggested several indicator-based algorithms, a trend that started in evolutionary multi-objective algorithm research [49]. Due to the importance of diversity for set-based optimization, recently developed indicator-based algorithms tend to include diversity, either as a separate indicator, see e.g. [14], or as an integral part of an indicator, see e.g. [40].

Although there is no consensus on how to best capture and use the concept of diversity in optimization, some robust definitions of diversity and measures are pointed out in [20, 43]. A diversity index measuring the number of different potential solutions and also the spread of solutions is recommended.

Different diversity indicators were proposed in the context of diversity-oriented search. Ulrich et al. [41] suggested to chose indicators from bio-diversity. In this field the Weitzmann diversity [43] and the Solow Polasky indicator [34] are common diversity measures. While the Weitzmann indicator [43] is motivated by phylogenetic trees with maximum parsimony and has exponential time complexity, the Solow Polasky indicator is motivated by a utilitarian model of species conservation and its computation can be accomplished in polynomial time. Due to its higher efficiency the latter indicator is favored by Ulrich et al. [41]. Even faster are indicators based on simple statistics on gaps between nearest neighbors [14, 26], although they can only provide comparisons among populations of the same size.

Ulrich et al. [39] emphasized the importance of decision space diversity by suggesting diversity-optimizing single objective (NOAH) and multi-objective (DIOP) algorithms, see [39, 41], respectively, as well as an algorithm that integrates diversity within the hypervolume indicator, see DIVA algorithm in [40].

When searching for *level set* approximations (e.g. for approximating an implicitly defined manifold), set-proximity indicators are required, which often strongly correlate with diversity indicators. Several diversity-based indicators were compared in [14], and the Hausdorff distance-based indicator was suggested for level set approximation within the Evolutionary Level Set Approximation (ELSA) algorithm. Indicators for multimodal optimization are discussed in [26].

An alternative approach to balance convergence and diversity in evolutionary algorithms was proposed in [44]. It stems from portfolio-optimization. It is well-known in strategic decision making and financial management, and it is based on the idea of composing portfolios of assets, which have a high potential of high return in future and are of lowest possible risk of failing (for the same reasons). The latter is related to how assets differ from each other. Diversity of assets selected in the same portfolio is shown to reduce potential risk for the portfolio of assets as a whole. By analogy the population of individuals in evolutionary algorithms can be selected as a portfolio of diverse, highly performing individuals. It can be formulated as a bi-objective optimization problem with two objectives: maximizing potential return of a portfolio as a whole and minimizing risk (or, in other words, maximizing diversity). These two objectives can also be combined into a single indicator, e.g. Sharpe ratio. This indicator shows how return compensates the risk taken. Traditionally, similarity of assets can be measured by covariance of their individual returns, which could be evaluated in a probabilistic sense taking into account their current performance. Then individuals with low (preferably negative) covariance are selected in the portfolio. Such selection can be done at both the parental and the environmental selection phases of evolutionary algorithms, and the latter can even be combined with preserving solutions in the archive as in the Portfolio Optimization Selection Evolutionary Algorithm (POSEA) [44]. Testing the approach on some benchmarks shows its efficiency when compared to the state-of-the-art indicator-based evolutionary algorithms and its potential for many objective optimization application due to the fact that its complexity is independent of the number of objectives.

6   Taxonomy and Ontology

In artificial intelligence, ontologies are used to formally represent a knowledge domain [16]. All instances, attributes and classes in the universum of discourse are represented as well as relations between these concepts and instances. As opposed to simple taxonomies or hierarchical descriptions of a field, ontologies allow to model in parallel different concepts and relate them with rules. An advantage of representing the survey of the domain of diversity-oriented optimization by means of an ontology is that besides the formal representation of classes and relations between them, also predicate logic rules can be specified that will help the user to classify algorithms correctly and find related work. Moreover, graphical representations of the ontology allow for a quick assessment of the research activities within this field and how they are related with each other.

Preliminary proposals for this taxonomy and ontology are provided in Figs. 3 and 4, respectively, and have been developed with the Protégé ontology editor [37].

Most relevant classes, concepts and relations presented graphically in the ontology were contextualized and described in detail in the previous sections.

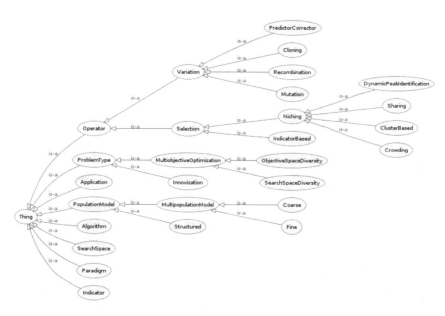

Fig. 3   Diversity oriented optimization taxonomy

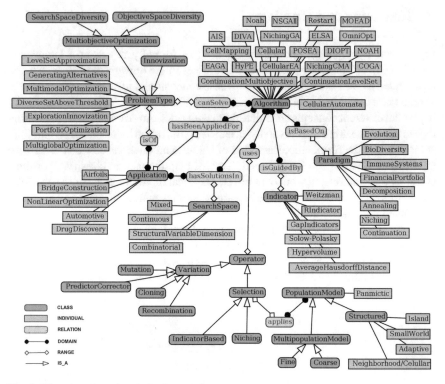

Fig. 4 Diversity oriented optimization ontology

7 Conclusion and Future Research

The overview proposed in this work aimed at capturing the development directions of diversity-oriented optimization algorithms. To represent the domain of diversity-oriented search in a systematic way, a diversity-oriented optimization taxonomy was established following an ontology discussion. In the nearest future, we aim at extending this ontology with various concepts, definitions and relations of diversity indicators, algorithms, and integrate diversity-related theoretical results into the survey.

In some application domains, the need for generating diverse solution sets has been particularly stressed and diversity-oriented optimization was successfully applied, for instance, in discrete design optimization in the car industry [38], truss bridge design and optimization [41], drug discovery [42, 45], quantum control [30], and space mission design [29]. It is expected that this is just the beginning and diversity oriented optimization will become increasingly important to design high performce and user friendly designs and search tools.

References

1. Alba, E., Dorronsoro, B.: Cellular Genetic Algorithms, Operations Research/Computer Science Interfaces, vol. 42. Springer, Heidelberg (2008)
2. Beekman, M., Sword, G.A., Simpson, S.J.: Biological foundations of swarm intelligence. In: Blum, C., Merkle, D. (eds.) Swarm Intelligence, Natural Computing Series, pp. 3–41. Springer, Heidelberg (2008)
3. Branke, J., Kaußler, T., Smidt, C., Schmeck, H.: A multi-population approach to dynamic optimization problems. In: Parmee, I. (ed.) Proceedings of the 4th International Conference on Adaptive Computing in Design and Manufacture (ACDM'2000, Plymouth, UK, April 26–28, 2000), pp. 299–307. Springer, Heidelberg (2000)
4. Burnet, F.: Clonal selection and after. In: Bell, G.I., Perelson, A.S., Pimbley Jr., G.H. (eds.) Theoretical Immunology, pp. 63–85. Marcel Dekker Inc., New York (1978)
5. de Castro, L., Von Zuben, F.: Learning and optimization using the clonal selection principle. Evol. Comput. IEEE Trans. **6**(3), 239–251 (2002)
6. Ceriani, L., Verme, P.: The origins of the Gini index: Extracts from Variabilità e Mutabilità (1912) by Corrado Gini. J. Econ. Inequal. **10**(3), 421–443 (2012)
7. Coelho, G.P., von Zuben, F.J.: Omni-aiNet: An immune-inspired approach for omni optimization. In: Bersini, H., Carneiro, J. (eds.) Proceeding of the 5th International Conference on Artificial Immune Systems (ICARIS, Oeiras, Portugal, September 4–6, 2006). Lecture Notes in Computer Science, vol. 4163, pp. 294–308. Springer, Heidelberg (2006)
8. Coelho, G.P., Von Zuben, F.J.: A concentration-based artificial immune network for multi-objective optimization. In: Takahashi, R.H.C., et al. (eds.) Proceedings of the 6th International Conference on Evolutionary Multi-Criterion Optimization (EMO 2011, Ouro Preto, Brazil, April 5–8, 2011). Lecture Notes in Computer Science, vol. 6576, pp. 343–357. Springer, Heidelberg (2011)
9. de Castro, L., Timmis, J.: An artificial immune network for multimodal function optimization. In: Proceedings of the 2002 Congress on Evolutionary Computation, (CEC 2002, Honolulu, Hawaii, USA, May 12–17, 2002), vol. 1, pp. 699–704. IEEE (2002)
10. Deb, K.: Innovization: Discovery of innovative solution principles using multi-objective optimization. In: Purshouse, R., et al. (eds.) Proceedings of the 7th International Conference on Evolutionary Multi-Criterion Optimization (EMO 2013, Sheffield, UK, March 19–22, 2013), pp. 4–5. Springer, Heidelberg (2013)
11. Deb, K., Srinivasan, A.: Innovization: Innovating design principles through optimization. In: Cattolico, M. (ed.) Proceedings of the 8th Annual Conference on Genetic and Evolutionary Computation (GECCO'06, Seattle, WA, USA, July 08–12, 2006), pp. 1629–1636. ACM, New York (2006)
12. Deb, K., Srinivasan, A.: Innovization: Discovery of innovative design principles through multiobjective evolutionary optimization. In: Knowles, J., et al. (eds.) Multiobjective Problem Solving from Nature, Natural Computing Series, pp. 243–262. Springer, Heidelberg (2008)
13. Deb, K., Tiwari, S.: Omni-optimizer: A generic evolutionary algorithm for single and multi-objective optimization. Eur. J. Oper. Res. **185**(3), 1062–1087 (2008)
14. Emmerich, M.T., Deutz, A.H., Kruisselbrink, J.: On quality indicators for black-box level set approximation. In: Tantar, E., et al. (eds.) EVOLVE - A bridge between Probability, Set Oriented Numerics and Evolutionary Computation, Studies in Computational Intelligence, vol. 447, pp. 157–185. Springer, Heidelberg (2012)
15. Ghosh, J.B.: Computational aspects of the maximum diversity problem. Oper. Res. Lett. **19**(4), 175–181 (1996)
16. Gruber, T.R.: A translation approach to portable ontology specifications. Knowl. Acquis. **5**(2), 199–220 (1993)
17. Guiasu, R.C., Guiasu, S.: The Rich-Gini-Simpson quadratic index of biodiversity. Nat. Sci. **2**(10), 1130–1137 (2010)
18. Jerne, N.K.: Towards a network theory of the immune system. Ann. Immunol. **125C**, 373–389 (1974)

19. Jin, Y., Branke, J.: Evolutionary optimization in uncertain environments-a survey. IEEE Trans. Evol. Comput. **9**(3), 303–317 (2005)
20. Jost, L.: Entropy and diversity. OIKOS **113**(2), 363–375 (2006)
21. Knowles, J.: Closed-loop evolutionary multiobjective optimization. IEEE Comput. Intell. Mag. **4**(3), 77–91 (2009)
22. Knowles, J., Corne, D.: The Pareto archived evolution strategy: a new baseline algorithm for Pareto multiobjective optimisation. In: P.J. Angeline, et al. (eds.) Proceedings of the 1999 Congress on Evolutionary Computation (CEC 99, Washington, USA, July 6–9, 1999), vol. 1, pp. 98–105. IEEE, New Jersey (1999)
23. Laumanns, M., Rudolph, G., Schwefel, H.P.: A spatial predator-prey approach to multi-objective optimization: A preliminary study. In: Eiben, A.E., et al. (eds.) Proceedings of the 5th International Conference on Parallel Problem Solving from Nature (PPSN V, Amsterdam, The Netherlands, September 27–30, 1998). Lecture Notes in Computer Science, vol. 1498, pp. 241–249. Springer, Heidelberg (1998)
24. Parmee, I.C., Bonham, C.R.: Towards the support of innovative conceptual design through interactive designer/evolutionary computing strategies. AI EDAM **14**(1), 3–16 (2000)
25. Pauling, L.: The Nature of the Chemical Bond and the Structure of Molecules and Crystals: An Introduction to Modern Structural Chemistry, vol. 18, 3d edn. Cornell University Press, Ithaca (1960)
26. Preuß, M., Wessing, S.: Measuring multimodal optimization solution sets with a view to mul-tiobjective techniques. In: Emmerich, M.T., et al. (eds.) Proceedings of the 4th International Conference: EVOLVE-A Bridge between Probability, Set Oriented Numerics, and Evolution-ary Computation (EVOLVE 2013, Leiden, The Netherlands, July 10–13, 2013), Advances in Intelligent Systems and Computing, vol. 227, pp. 123–137. Springer, Heidelberg (2013)
27. Reehuis, E., Kruisselbrink, J., Olhofer, M., Graening, L., Sendhoff, B., Bäck, T.: Model-guided evolution strategies for dynamically balancing exploration and exploitation. In: Hao, J., et al. (eds.) Proceedings of the 10th International Conference on Artificial Evolution, (EA 2011, Angers, France, October 24–26, 2011), pp. 306–317. Springer, Heidelberg (2011)
28. Schönemann, L., Emmerich, M.T., Preuß, M.: On the extinction of evolutionary algorithm subpopulations on multimodal landscapes. Informatica (Slowenien) **28**(4), 345–351 (2004)
29. Schütze, O., Vasile, M.: Coello Coello, C.A.: Approximate solutions in space mission design. In: Proceedings of the 10th International Conference on Parallel Problem Solving from Nature (PPSN X. Dortmund, Germany, September 13–17, 2008). Lecture Notes in Computer Science, vol. 5199, pp. 805–814. Springer, Berlin (2008)
30. Shir, O., Beltrani, V., Bäck, T., Rabitz, H., Vrakking, M.: On the diversity of multiple optimal controls for quantum systems. J. Phys. B At. Mol. Opt. Phys. **41**(7), (2008)
31. Shir, O., Preuß, M., Naujoks, B., Emmerich, M.: Enhancing decision space diversity in evo-lutionary multiobjective algorithms. Evolutionary Multi-Criterion Optimization. Studies in Computational Intelligence, pp. 95–109. Springer, Heidelberg (2009)
32. Shir, O.M.: Niching in evolutionary algorithms. In: Rozenberg, G., Bäck, T., Kok, J.N. (eds.) Handbook of Natural Computing: Theory, Experiments, and Applications, pp. 1035–1069. Springer, Heidelberg (2012)
33. Simpson, E.H.: Measurement of diversity. Nature **163**(4148), 688 (1949)
34. Solow, A., Polasky, S., Broadus, J.: On the measurement of biological diversity. J. Environ. Econ. Manag. **24**(1), 60–68 (1993)
35. Solow, A.R., Polasky, S.: Measuring biological diversity. Environ. Ecol. Stat. **1**(2), 95–107 (1994)
36. Stoean, C., Preuß, M., Stoean, R., Dumitrescu, D.: Multimodal optimization by means of a topological species conservation algorithm. IEEE Trans. Evol. Comput. **14**(6), 842–864 (2010)
37. Tudorache, T., Nyulas, C., Noy, N.F., Musen, M.A.: WebProtégé: A collaborative ontology editor and knowledge acquisition tool for the web. Semant. web **4**(1), 89–99 (2013)
38. Ulrich, T.: Exploring structural diversity in evolutionary algorithms. Ph.D. thesis, ETH Zurich, TIK Institut für Technische Informatik und Kommunikationsnetze (2012)

39. Ulrich, T., Bader, J., Thiele, L.: Defining and optimizing indicator-based diversity measures in multiobjective search. In: Schaefer, R., et al. (eds.) Proceedings of the 11th International Conference on Parallel Problem Solving from Nature: Part I (PPSN XI, Krakow, Poland, September 11–15, 2010), pp. 707–717. Springer, Heidelberg (2010)
40. Ulrich, T., Bader, J., Zitzler, E.: Integrating decision space diversity into hypervolume-based multiobjective search. In: Pelikan, M., Branke, J. (eds.) Proceedings of the 12th Annual Conference on Genetic and Evolutionary Computation (GECCO'10, Portland, USA, July 07–11, 2010), pp. 455–462. ACM, New York (2010)
41. Ulrich, T., Thiele, L.: Maximizing population diversity in single-objective optimization. In: Krasnogor, N., Lanzi, P.L. (eds.) Proceedings of the 13th Annual Conference on Genetic and Evolutionary Computation (GECCO '11, Dublin, Ireland, July 12–16, 2011), pp. 641–648. ACM, New York (2011)
42. van der Horst, E., Marqués-Gallego, P., Mulder-Krieger, T., van Veldhoven, J., Kruisselbrink, J., Aleman, A., Emmerich, M.T., Brussee, J., Bender, A.: IJzerman, A.P.: Multi-objective evolutionary design of adenosine receptor ligands. J. Chem. Inf. Model. **52**(7), 1713–1721 (2012)
43. Weitzman, M.L.: On diversity. Q. J. Econ. **107**(2), 363–405 (1992)
44. Yevseyeva, I., Guerreiro, A.P., Emmerich, M.T., Fonseca, C.M.: A portfolio optimization approach to selection in multiobjective evolutionary algorithms. In: Bartz-Beielstein, T., et al. (eds.) Proceedings of the 13th International Conference on Parallel Problem Solving from Nature (PPSN XIII, Ljubljana, Slovenia, September 13–17, 2014). Lecture Notes in Computer Science, vol. 8672, pp. 672–681. Springer, Heidelberg (2014)
45. Yevseyeva, I., Lenselink, E.B., de Vries, A., Ijzerman, A.P., Deutz, A.H., Emmerich, M.T.: Multiobjective portfolio optimization for drug discovery using deterministic and stochastic methods. In: M.J. Geiger (ed.) Abstracts of the 23d International Conference on Multicriteria Decision Making (MCDM 2015 - Bridging Disciplines, Hamburg, Germany, August 2–7 (2015)
46. Zadorojniy, A., Masin, M., Greenberg, L., Shir, O.M., Zeidner, L.: Algorithms for finding maximum diversity of design variables in multi-objective optimization. Procedia Comput. Sci. **8**, 171–176 (2012)
47. Zechman, E., Ranjithan, S.: An evolutionary algorithm to generate alternatives (EAGA) for engineering optimization problems. Eng. Optim. **36**(5), 539–553 (2004)
48. Zechman, E., Ranjithan, S.: Evolutionary computation-based methods for characterizing contaminant sources in a water distribution system. J. Water Res. Planning Manag. **135**(5), 334–343 (2009)
49. Zitzler, E., Künzli, S.: Indicator-based selection in multiobjective search. In: Yao, X. (ed.) Proceedings of the 8th International Conference on Parallel Problem Solving from Nature (PPSN VIII, Birmingham, UK, September 18–22, 2004), pp. 832–842. Springer-Verlag, Berlin, Heidelberg (2004)

Global Multi-objective Optimization by Means of Cell Mapping Techniques

Carlos Hernández, Oliver Schütze and Jian-Qiao Sun

Abstract Multi-objective optimization problems (MOPs) arise in many fields in engineering. In this chapter we argue that adaptation of cell mapping techniques, originally designed for the global analysis of dynamical systems, are well-suited for the thorough analysis of low-dimensional MOPs. Algorithms of this kind deliver an approximation of the set of global solutions, the Pareto set, as well as the set of locally optimal and nearly optimal solutions in one run of the algorithm which may significantly improve the underlying decision making process. We underline the statements on some illustrative examples and present comparisons to other algorithms.

1 Introduction

In many applications the problem arises that several objectives have to be optimized concurrently. For instance, two important objectives in many space mission design problems are the time of flight and the cost for a mission to a certain destination [1]. One important characteristic of such a *multi-objective optimization problem* (MOP) is that its solution set, the *Pareto set*, does typically not consist of a singleton but forms a $(k-1)$-dimensional object, where k is the number of objectives involved in the MOP. The computation of Pareto sets thus represents in general a challenge. Even more, in certain applications one may be interested in approximate solutions that may allow the decision maker (DM) to find alternative or backup solutions to a given problem. As for the above space mission design problem this could be a trajectory that is slightly more costly and has a slightly longer time of flight than an

C. Hernández (✉) · O. Schütze
Computer Science Department, CINVESTAV-IPN, Mexico City, Mexico
e-mail: chernandez@computacion.cs.cinvestav.mx

O. Schütze
e-mail: schuetze@cs.cinvestav.mx

J.-Q. Sun
School of Engineering, University of California Merced, Merced, USA
e-mail: jsun3@ucmerced.edu

© Springer International Publishing AG 2017

M. Emmerich et al. (eds.), *EVOLVE – A Bridge Between Probability,
Set Oriented Numerics and Evolutionary Computation VII*,
Studies in Computational Intelligence 662, DOI 10.1007/978-3-319-49325-1_2

optimal one but offers in turn a different realization of the problem (see [2] for such examples where the launch date has been considered as influential for the related decision making process). The set of these approximate solutions even forms an n-dimensional set, where n is the number of decision variables involved in the model.

There exist so far many techniques for the numerical treatment of MOPs. Most of these works focus on the detection of one or several optimal solutions, while the consideration of approximate solutions is relatively scarce, probably due to the dimension of the solution set. There exists, for instance, a variety of point-wise iterative search procedures that are capable of generating a sequence of points moving either toward or along the Pareto set (e.g., [3–7] and references therein). These local methods, however, are not universally applicable due to the fact that the solution set is not a singleton as well as some possible characteristics of the model such as multi-modality and disconnectedness of the domain and/or the set of interest. A possible alternative is given by set oriented methods that are of global nature but, in turn, applicable to lower dimensional problems. Among them, specialized evolutionary strategies have caught the interest of many researchers in the recent past (e.g., [8–10]) since algorithms of this kind are very robust, applicable to a broad class of problems, and deliver a finite size approximation of the set of interest in one single run. Another class of set oriented methods are subdivision techniques [11–13] that start by considering an n-dimensional box that contains the domain of the MOP. This box gets subdivided into a set of smaller boxes, and according to certain conditions it is decided which box could contain a part of the set of interest and is thus suited for further investigation. The other, unpromising boxes, are discarded from the collection. This process, subdivision and selection, is performed on the current box collection until the desired granularity of the boxes is reached. This way, a tight covering of the Pareto set is obtained.

In this chapter, we argue that *cell mapping techniques* are in particular advantageous for the thorough investigation of low dimensional problems. Such problems occur such problems occur, for instance, in optimal control [14–17]. Cell mapping techniques were first introduced in [18] for global analysis of nonlinear dynamical systems. They transform classical point-to-point dynamics into a cell-to-cell mapping by discretizing both phase space and the integration time. In particular the phase space discretization bounds the method to a small number of variables that can be considered (say, $n < 10$), but this global analysis offers in turn much more information than other methods. In the context of multi-objective optimization this is in particular the extended set of options that can be offered to the DM after analyzing the model. There are first of all the Pareto set and the set of approximate solutions as motivated above. In particular if there exist several possibilities to obtain the same optimal or nearly optimal performance, other methods have problems to detect them all since the notion of dominance is defined in objective space (and thus, typically only one of these solutions is detected). Further, the entire set of local optima can be identified that also serve as potential backup solutions [2] and that are interesting for landscape analysis [19]. It is important to note that the relevant information about all these sets of interest is available after one single run of the algorithm (together with an ex post analysis of the obtained data).

In this work we will investigate adaptations of the cell mapping techniques to the context of multi-objective optimization where we will concentrate on the computation of optimal and nearly optimal solutions of a given MOP. A preliminary study of approximate solutions in the sense of Loridan [20] by means of cell mapping can be found in [21]. Further, applications of the method to the design of optimal feedback control are presented in [15, 16].

The remainder of this chapter is organized as follows: in Sect. 2, we state the notations and some background required for the understanding of the chapter. In Sect. 3, we state the cell mapping techniques for MOPs. In Sect. 4, we will present some numerical results. Finally, in Sect. 5, we conclude and will give some paths for future work.

2 Notations and Background

In the following we consider continuous MOPs

$$\min_{x \in Q} F(x), \tag{MOP}$$

where $Q \subset \mathbb{R}^n$ is the domain of the problem and F is defined as the vector of the objective functions $F : Q \to \mathbb{R}^k$, $F(x) = (f_1(x), \ldots, f_k(x))^T$, and where each objective $f_i : \mathbb{R}^n \to \mathbb{R}$ is (for simplicity) sufficiently smooth.

The optimality of a MOP is defined by the concept of *dominance* [22]: a vector $v \in \mathbb{R}^k$ is less than $w \in \mathbb{R}^k$ ($v <_p w$), if $v_i < w_i$ for all $i \in 1, \ldots, k$. The relation $\leq_p$ is defined analogously. Then, a vector $y \in Q$ is *dominated* by a vector $x \in Q$ (in short: $x \prec y$) with respect to (MOP) if and only if $f_i(x) \leq f_i(y)$, $i = 1, \ldots, k$, and there exists an index j such that $f_j(x) < f_j(y)$, else y is non-dominated by x. A point $x \in Q$ is called *(Pareto) optimal* or a *Pareto point* if and only if there is no $y \in Q$ which dominates x. The set of all Pareto optimal solutions P_Q is called the *Pareto set*, and its image $F(P_Q)$ the *Pareto front*. Both sets typically form a $(k-1)$-dimensional object.

To define the set of approximate solutions we need the following definition.

Definition 1 ([20, 23]) Let $\varepsilon = (\varepsilon_1, \ldots, \varepsilon_k) \in \mathbb{R}_+^k$ and $x, y \in Q$.

(a) x is said to ε-dominate y ($x \prec_\varepsilon y$) with respect to (MOP) if and only if $F(x) - \varepsilon \leq_p F(y)$ and $F(x) - \varepsilon \neq F(y)$.
(b) x is said to $-\varepsilon$-dominate y ($x \prec_{-\varepsilon} y$) with respect to (MOP) if and only if $F(x) + \varepsilon \leq_p F(y)$ and $F(x) + \varepsilon \neq F(y)$.

The notion of $-\varepsilon$-dominance can be used to define our set of interest.

Definition 2 ([23]) Denote by $P_{Q,\varepsilon}$ the set of points in $Q \subset \mathbb{R}^n$ that are not $-\varepsilon$-dominated by any other point in Q, i.e.,

$$P_{Q,\varepsilon} := \{x \in Q | \nexists y \in Q : y \prec_{-\varepsilon} x\}. \tag{1}$$

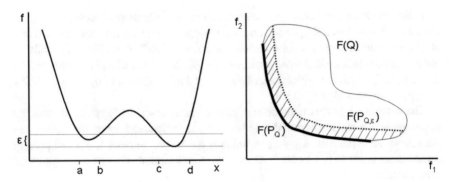

Fig. 1 Two different examples for sets $P_{Q,\varepsilon}$. At the *left*, we show the case for $k = 1$ and in parameter space with $P_{Q,\varepsilon} = [a, b] \cup [c, d]$. Note that the image solutions $f([a, b])$ are nearly optimal (measured in objective space), but that the entire interval $[a, b]$ is not 'near' to the optimal solution which is located within $[c, d]$. At the *right*, we show an example for $k = 2$ in image space, $F(P_{Q,\varepsilon})$ is the approximate Pareto front (taken from [23])

The set $P_{Q,\varepsilon}$ contains all ε-efficient solutions, i.e., solutions which are optimal up to a given (small) value of ε. See Fig. 1 for two examples.

In [23, 24] several archiving techniques were proposed. In this work, we focus in *ArchiveUpdateP$_{Q,\varepsilon}$*. The archiver guarantees convergence under certain assumptions on the MOP and the generation process (see for more details [23, 24]). Algorithm 1 show the realization of the archiver.

Algorithm 1 $A := ArchiveUpdateP_{Q,\varepsilon}\ (P, A_0, \varepsilon)$

Require: population P, archive A_0
Ensure: updated archive A
1: $A := A_0$
2: **for all** $p \in P$ **do**
3:    **if** $\nexists a \in A : a \prec_{-\varepsilon} p$ **then**
4:      $A := A \cup \{p\}$
5:    **end if**
6:    **for all** $a \in A$ **do**
7:      **if** $p \prec_{-\varepsilon} a$ **then**
8:        $A := A \backslash \{a\}$
9:      **end if**
10:   **end for**
11: **end for**

The cell mapping method was originally proposed by Hsu [18, 25] for global analysis of nonlinear dynamical systems in the *state space*. The cell mapping methods have been extensively studied, which lead to, the simple cell mapping, the generalized cell mapping [25], the interpolated cell mapping [26], the adjoining cell mapping [27, 28], the hybrid cell mapping [29], among others. The cell mapping methods have been applied to optimal control problems of deterministic and stochastic dynamical systems [14, 17, 30, 31]. In [28], the cell mapping techniques where combined with dynamical systems theory in order to find all solutions to a system of nonlinear algebraic equations.

The cell mapping methods transform the point-to-point dynamics into a cell-to-cell mapping by discretizing both phase space and the integration time. The simple cell mapping (SCM) offers an effective approach to investigate global response properties of the system. The cell mapping with a finite number of cells in the computational domain will eventually lead to closed groups of cells of the period equal to the number of cells in the group. The periodic cells represent approximate invariant sets, which can be periodic motion and stable attractors of the system. The rest of the cells form the domains of attraction of the invariant sets. For more discussions on the cell mapping methods, their properties and computational algorithms, the reader is referred to the book by Hsu [25].

3 Global Analysis of Dynamical Systems

In this section, we first define a dynamical system, and further on the solution of a dynamical system. We also present the concept of the domain of attraction and finally we look into the simple cell mapping method that was proposed to perform a global analysis of a given dynamical system.

3.1 Dynamical Systems

Definition 3 (*Dynamical System*) A dynamical system [25] can be considered to be a model describing the temporal evolution of a system and it is defined as follows:

$$\dot{x} = G(x),$$

where x is a n-dimensional vector and $G : \mathbb{R}^n \to \mathbb{R}^n$ is, in general, a nonlinear vector function. The evolution of such a dynamical system can be described by a function of the form:

$$x_{m+1} = G(x(m), \mu), \tag{2}$$

where x is a n-dimensional vector, m denotes the mapping step, μ is a parameter vector, and G is a general nonlinear vector function. In this case ordinary differential equations can be used to describe the dynamical systems. These are defined as follows:

$$\dot{x} = F(x, t, \mu); \ x \in \mathbb{R}^n, \ t \in R, \ \mu \in \mathbb{R}^l,$$

where x is a n-dimensional state vector, t is the time variable, μ is a l-dimensional parameter vector, and F is a vector-valued function of x, t and μ.

Definition 4 (*Fixed point*) When the evolution of a dynamical system is made, one may find a point that satisfies the following:

$$x^* = G(x^*, \mu).$$

In this case, x^* is called a fixed point of Eq. (2).

Definition 5 (*Periodic group*) A periodic solution of Eq. (2) of period l is a sequence of l distinct points $x^*(j)$, $j = 1, 2, \ldots, l$ such that

$$x^*(o + 1) = G^o(x^*(1), \mu), \ o = 1, 2, \ldots, l - 1,$$
$$x^*(1) = G^l(x^*(1), \mu). \tag{3}$$

We say that there exists a periodic solution of period l. Any of the points $x^*(j)$, $j = 1, 2, \ldots, l$, is called a periodic point of period l. One can see that a fixed point is a periodic solution with $l = 1$.

Definition 6 (*Domain of attraction*) We say $x^*(j)$ is an attractor if there exists a neighborhood U of $x^*(j)$ such that for every open set $V \supset x^*(j)$ there is a $N \in \mathbb{N}$ such that $f^j(U) \subset V$ for all $j \geq N$. Hence, we can restrict ourselves to the closed invariant set $x^*(j)$, and in this case we obtain

$$x^*(j) = \bigcap_{j \in N} G^j(U).$$

Thus, we can say that all the points in U are attracted by $x^*(j)$ (under iteration of G), and U is called basin of attraction of $x^*(j)$. If $U = \mathbb{R}^n$, then $x^*(j)$ is called the global attractor.

Several kinds of attractors exists, however, only the ones formed by the set of periodic solutions will be considered in this work.

3.2   Simple Cell Mapping

In this section, we present the simple cell mapping [25], which is useful to compute global attractors and domains of attraction of a given dynamical system.

SCM does not consider the state space to be continuous but rather as a collection of state cells, with each cell being taken as a state entity. Because of this, now we need to introduce some basic concepts regarding the new model.

Definition 7 (*Cell state space*) A n dimensional cell space S [25] is a space whose elements are n-tuples of integers, and each element is called a cell vector or simply a cell, and is denoted by z.

The simplest way to obtain a cell structure over a given Euclidean state space is to construct a cell structure consisting of rectangular parallelepipeds of uniform size.

Definition 8 (*Cell functions*) Let S be the cell state space for a dynamical system and let the discrete time evolution process of the system be such that each cell in a region of interest $S_0 \subset S$ has a single image cell after one mapping step. Such an evolution process is called simple cell mapping (SCM)

$$z(n + 1) = C(z(n), \mu), z \in \mathbb{Z}^N, \mu \in \mathbb{R}^l, \tag{4}$$

where $C : \mathbb{Z}^N \times \mathbb{R}^l \to \mathbb{Z}^N$, and μ is a l-dimensional parameter.

Definition 9 (*Periodic group*) A cell z^* which satisfies $z^* = C(z*)$ is said to be an equilibrium cell of the system. Let C^m denote the cell mapping C applied m times with C^0 understood to be the identity mapping. A sequence of l distinct cells $z^*(j), j \in l$, which satisfies

$$z^*(m + 1) = C^m(z^*(1)), m \in l - 1, z^*(1) = C^l(z^*(1)), \tag{5}$$

is said to constitute a periodic group or P-Group of period l and each of its elements $z^*(j)$ a periodic cell of period l. One can see that an equilibrium cell is a $l = 1$ periodic group.

Definition 10 (*Domains of attraction*) A cell z is said to be r steps away from a periodic group if r is the minimum positive integer such that $C^r(z) = z^*(j)$, where $z^*(j)$ is one of the cells of that periodic group.

The set of all cells, which are r steps or less removed from a periodic group is called the r-step domain of attraction for that periodic group. The total domain of attraction of a periodic group is its r-step domain of attraction with $r \to \infty$.

The main idea of this method is based on the fact that the representation of the numbers in a computer is finite. A number does not only represent the number represented by its digits, but also an infinite neighborhood of numbers given by the precision of the machine. This does not allow to assume variables to be continuous, due to rounding errors and for this reason it is possible to consider the space as small hypercubes whose size is given by the machine precision.

The cell mapping approach [25] proposes to increase this discretization by dividing the state space into bigger hypercubes. The evolution of the dynamical system is then reduced to a new function, which is not defined in $\mathbb{R}^n$, but rather on the cell space. In this case we restrict ourselves to functions that are strictly deterministically defined. For this case, we have the so-called simple cell mapping method, which is effective to obtain the attractors and basins of attraction of a dynamical system.

The SCM method uses some sets in order to capture the global properties of a cell, which we describe in the following:

- Group motion number (Gr): the group number uniquely identifies a periodic motion; it is assigned to every periodic cell of that periodic motion and also to every cell in the domain of attraction. The group numbers, which are positive integers, can be assigned sequentially.

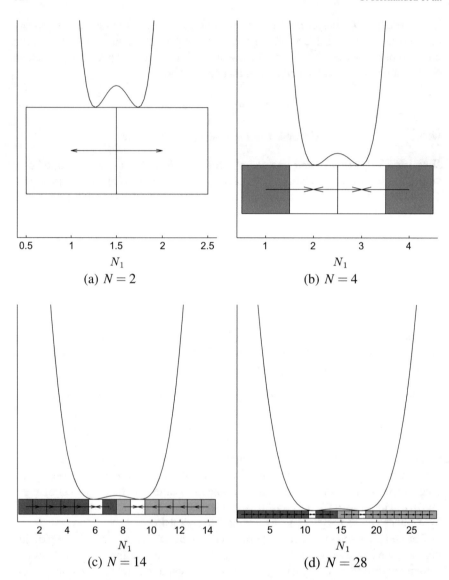

Fig. 2 Numerical result of the SCM on Eq. (6) with different grid size, **a** $N = 2$; **b** $N = 4$; **c** $N = 14$; **d** $N = 28$. The *white cells* represents the optimal solutions. The cells with the same color belong to the same domain of attraction (for those cells their mapping end in the same cell). The *arrows* represent the cell mapping. Finally, the *black curve* is the graphic representation of the problem

- Period (Pe): defines the period of each periodic motion.
- Number of steps to a P-group (St): used to indicate how many steps it takes to map this cell into a periodic cell.

According to the previous discussion, the algorithm works as follows: until all cells are processed, the value of the group motion indicates the state of the current cell and it also points out the corresponding actions to the cell.

- A value of $Gr(cell) = 0$ means that the cell has not been processed. Hence, the state of the cell changes to "under process" and then, we follow the dynamical system to the next cell.
- A value of $Gr(cell) = -1$ means that the cell is under process, which means we have found a periodic group and we can compute the global properties of the current periodic motion.
- A value $Gr(cell) > 0$ means that the cell has already been processed. Hence we found a previous periodic motion along with its global properties which can be used to complete the information of the cells under process.

The cell mapping methods have been applied to optimal control problems of deterministic and stochastic dynamic systems [30–32]. Other interesting applications of the cell mapping methods include optimal space craft momentum unloading [33], single and multiple manipulators of robots [34], optimum trajectory planning in robotic systems by [35], and tracking control of the read-write head of computer hard disks [36].

Now, we present an application of the SCM on a simple example. We consider the following SOP:

$$\min_x f = 4x^3 - 2x, \tag{6}$$

where $f : \mathbb{R} \to \mathbb{R}$ and $x \in \mathbb{R}$. For this problem, we have two optimal points at $\frac{\sqrt{2}}{2}$ and $-\frac{\sqrt{2}}{2}$. Figure 2 shows the result for different values of N and $Q = [-3, 3]$. The figure shows the mapping from one cell to another one, until it reaches a periodic group. Further, it shows two different group motions and for $N = 14$ and $N = 28$, we can also see where the domain of attraction of $\frac{\sqrt{2}}{2}$ ends and the domain of attraction of $-\frac{\sqrt{2}}{2}$ begins.

4 Simple Cell Mapping Techniques for MOPs

The cell mapping methods are so far designed for the global analysis of general nonlinear dynamical systems. In the following, we will present adaptations to the SCM in order to handle multi-objective optimization problems.

4.1 The Algorithm

First, we need an appropriate dynamical system to run SCM. For this, we propose to utilize *descent directions*. A descent direction v at a point $x_0 \in Q$ is a direction in which all objectives improve, i.e., it holds

$$F(x_0 + tv) <_p F(x_0) \tag{7}$$

for all sufficiently small step sizes $t \in \mathbb{R}_+$. Such descent directions can be found e.g., in [37–40]. For the bi-objective problems presented in this chapter, we have used the following one.

Theorem 1 ([39]) *Let (MOP) be unconstrained and be defined by two differentiable functions. If $\nabla f_i(x_0) \neq 0$, for $i = 1, 2$ and for $x_0 \in \mathbb{R}^n$, then*

$$v(x_0) := -\left(\frac{\nabla f_1(x_0)}{||\nabla f_1(x_0)||} + \frac{\nabla f_2(x_0)}{||\nabla f_2(x_0)||} \right) \tag{8}$$

is a descent direction at x_0 of (MOP).

Using such a descent direction, the following dynamical system

$$\dot{x}(t) = v(x(t)) \tag{9}$$

can now be used since it defines a pressure toward the Pareto set/front of the MOP at hand: $v(x) = 0$ for every (locally) optimal point, and for all other points improvements can be found by integrating (9). Thus, the set of locally optimal Pareto points is contained in the global attractor of (9).

It remains to discretize the time (9), i.e., to define a 'suitable' step size t for the related discrete dynamical system

$$x_{i+1} = x_i + tv(x_i). \tag{10}$$

This is in general not an easy task as we have two conflicting aims. On the one hand, we would like to choose a step size t that lowers all objective functions as much as possible for a given direction v. On the other hand, it is desired to make this decision as cheap as possible in terms of computing time and number of function evaluations. One option is to use an inexact step size control as the one proposed in [38].

Here, we can take advantage of the particular setting of the multi-objective SCM. Most importantly, we have the size $h = (ub_i - lb_i)/N_i$ for $i = 1, \ldots, n$ of the cells and know that the initial point is the center of a cell. Using this, we already have a value for sufficient decrease. If there exists a $tv_i \geq \frac{h_i}{2}$, $i = 1, \ldots, n$, then we ensure that we leave the current cell, which is required for the SCM in case the cell does not contain a part of the Pareto set. Now, to decide if the step size t is accepted, we

Fig. 3 Illustration of the setting of the step size control problem for the SCM method

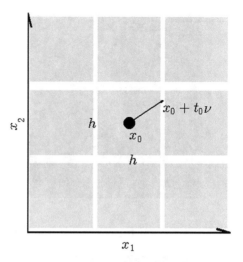

can use a dominance test. We are left with the choice of the initial step size t_0. In the current context, it is promising to compute the distance to the nearest neighbor cell given the descent direction v from the current cell center. Thus, we suggest to take (compare to Fig. 3):

$$t_0 = \max\left(\frac{h_i}{v_i}\right) + \varepsilon, \ \forall i | v_i \neq 0. \tag{11}$$

We used this approach in the present work, with good computational results. Alternatively, one could use a more sophisticated method such as the one presented in [28]. The authors of this work propose an adaptive integration scheme which either finds a neighboring cell or stays in the same cell.

In both cases, although a bigger value of t_0 may lead to a bigger decrease in the objective function, this value of t_0 is enough to leave the current cell and we have several advantages. We would lose less information since we would be moving to a neighbor cell. Also with this step size control we would be in the frontier between the current cell and its neighbor, thus if the step size t_0 is not accepted there is no need to use backtracking. Given that we would not be able to leave the current cell and also, since all cells are visited in the SCM method the advantages that bigger step sizes would have by going to an optimal solution with less function evaluations would be lost.

Inequality constraints are handled in the following (straightforward) way: if the center point x_i of cell i is violating any constraint, it will be discarded (i.e., mapped to the sink cell), else, the point will be mapped as described above. The inclusion of more sophisticated constraint handling techniques including the adequate treatment of equality constraints is the subject of ongoing study.

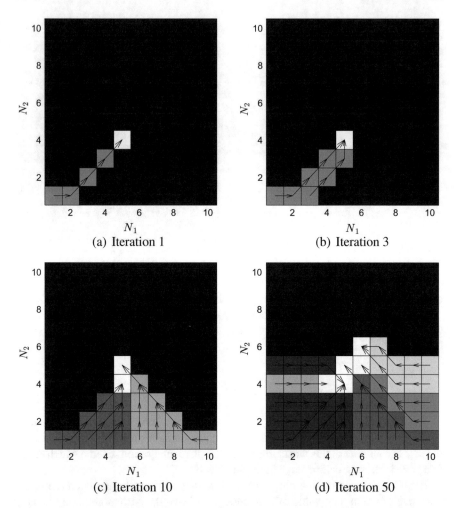

Fig. 4 Iteration of the SCM. The *white cells* represent the optimal solutions found so far. The *arrows* show the path from the starting cell to an optimal solution of the MOP. The *darker cells* represent unexplored regions

Algorithm 2 shows the pseudo code of the cell mapping technique for the treatment of MOPs that contains the above discussion. Figure 4 gives some insight into the behavior of the SCM using the MOP CONV2 (see Table 1) on a 10×10 grid. The figure shows the result of the SCM after 1, 3, 10 and 50 iterations in cell space. First, we look at the cell located in $(1, 1)$, which has been taken as the starting cell. Next, we can follow the mapping from this cell by following its arrow. These arrows are formed as follows: We take the center point of the current cell, then we apply the dynamical system (e.g., the descent direction method that we have chosen) on the center point and finally, with the new solution found, we compute to which cell it

Table 1 MOPs considered in this work

WITTING	$F(x) = (f_1(x), f_2(x))$, where: $f_1(x) = \frac{1}{2}(\sqrt{1+(x+y)^2} + \sqrt{1+(x-y)^2} + x - y) + \lambda \cdot e^{-(x-y)^2}$ $f_2(x) = \frac{1}{2}(\sqrt{1+(x+y)^2} + \sqrt{1+(x-y)^2} - x + y) + \lambda \cdot e^{-(x-y)^2}$	$-10 \le x_1 \le 10$ $-10 \le x_2 \le 10$
CONV2	$F(x) = (f_1(x), f_2(x))$, where: $f_1(x) = (x_1-1)^4 + (x_2-1)^2$ $f_2(x) = (x_1+1)^2 + (x_2+1)^2$	$-3 \le x_1 \le 3$ $-3 \le x_2 \le 3$
CONV3	$F(x) = (f_1(x), f_2(x))$, where: $f_1(x) = (x_1-1)^4 + (x_2-1)^2 + (x_3-1)^2$ $f_2(x) = (x_1+1)^2 + (x_2+1)^2 + (x_3+1)^2$ $f_3(x) = (x_1-1)^2 + (x_2+1)^4 + (x_3-1)^4$	$-3 \le x_1 \le 3$ $-3 \le x_2 \le 3$ $-3 \le x_3 \le 3$
RUDOLPH	$F(x) = (f_1(x), f_2(x))$, where: $f_1(x) = (x_1 - t_1(c+2a)+a)^2 + (x_2 - t_2 b)^2 + \delta_t$ $f_2(x) = (x_1 - t_1(c+2a)-a)^2 + (x_2 - t_2 b)^2 + \delta_t$ where $t_1 = \mathrm{sgn}(x_1)\min\left(\left\lceil \frac{|x_1|-a-c/2}{2a+c} \right\rceil, 1\right),$ $t_2 = \mathrm{sgn}(x_2)\min\left(\left\lceil \frac{|x_2|-b/2}{b} \right\rceil, 1\right),$ $\delta_t = \begin{cases} 0 & \text{for } t_1=0 \text{ and } t_2=0 \\ 0.01 & \text{for } t_1=-1 \text{ and } t_2=0 \\ 0.02 & \text{for } t_1=1 \text{ and } t_2=0 \\ 0.03 & \text{for } t_1=0 \text{ and } t_2=-1 \\ 0.04 & \text{for } t_1=-1 \text{ and } t_2=-1 \\ 0.05 & \text{for } t_1=1 \text{ and } t_2=-1 \\ 0.06 & \text{for } t_1=0 \text{ and } t_2=1 \\ 0.07 & \text{for } t_1=-1 \text{ and } t_2=1 \\ 0.08 & \text{for } t_1=1 \text{ and } t_2=1 \end{cases}$	$a = 0.5$ $b = 5$ $c = 5$ $-10 \le x_1 \le 10$ $-10 \le x_2 \le 10$
SSW	$F(x) = (f_1(x), f_2(x))$, where: $f_1(x) = \sum_{j=1}^{n} x_j,$ $f_2(x) = 1 - \prod_{j=1}^{n}(1 - w_j(x_j)),$ $w_j(z) = \begin{cases} 0.01 \cdot \exp(-(\frac{z}{20})^{2.5}) & \text{for } j=1,2 \\ 0.01 \cdot \exp(-\frac{z}{15}) & \text{for } j>2 \end{cases}$	$0 \le x_1 \le 40$ $0 \le x_2 \le 40$ $0 \le x_3 \le 40$
TANAKA	$F(x) = (f_1(x), f_2(x))$, where: $f_1(x) = x_1$ $f_2(x) = x_2$	$0 \le x_1 \le \pi$ $0 \le x_2 \le \pi$ $g_1(x) = -x_1^2 - x_2^2 + 1$ $+ 0.1\cos(16 atan(\frac{x_1}{x_2}))$ $g_2(x) = (x_1 - \frac{1}{2})^2$ $+ (x_2 - \frac{1}{2})^2 - 1/2$

belongs. In our example, the path is formed by the cells $(1, 1)$, $(2, 1)$, $(3, 2)$, $(4, 3)$, and $(5, 4)$. Cell $(5, 4)$ is an endpoint in this case, since there is not an arrow from this cell to another cell, which means we have a periodic group of 1. All the cells processed belong to the same domain of attraction and, therefore, they should have the same group number. Since, this is the first group motion discovered, we assign to it the group number 2 (the group number 1 is reserved for those periodic motions that go to the sink cell). Once we have the global properties of those cells, we have to choose a new starting cell. Since the cell $(2, 1)$ has already been processed, we skip it and continue with the cell $(3, 1)$. The mapping of this cell also finishes in the cell $(5, 4)$. Thus, this cell together with the new path should have the same group number as before (group number 2).

Then, we choose a new starting cell and continue until we finish processing all the cells. As we process the cells, we gather more information of the problem. For this example we have 8 periodic motions with the same number of optimal solutions.

After one run of the SCM the information of the sets of interest can be extracted. In the following, we will do this for optimal and nearly optimal solutions.

4.2 Computing the Pareto Set

Since the Pareto set of a MOP is contained in the global attractor of the dynamical system that is derived from a descent direction, all cells with periodic groups are at first point interesting. That is, such cells can potentially contain a part of the Pareto set. It is important to note that due to the properties of the dynamical system periodic groups with size larger than 1 should not appear, however, due to the discretization both in space and time exactly this happens (i.e., oscillations around Pareto optimal solutions can be observed leading to such periodic groups). Hence, we also consider those cells as candidates. The collection of those cells form the candidate set.

This collection can then be further investigated (e.g., via a more fine grain cell mapping or via subdivision techniques), or an approximation of the Pareto set can be directly determined via the center points of the boxes (e.g., via a non-dominance test). Technically speaking, we introduce a set called cPs (see Algorithm 2). Candidate optimal cells are thus those cells with $St = 0$ and $Gr \neq 1$. $St = 0$ means they are part of a periodic group and $Gr \neq 1$ ensures we do not add cells that map to the sink cell.

Algorithm 2 Simple cell mapping for MOPs

Require: MOP F, Dynamical system v, upper bound ub, lower bound lb, divisions per dimension N, cell size $h = (ub_i - lb_i)/N_i$ for $i = 1, \ldots, n$, Total number of cells $N_c = N_1 \times N_2 \times \ldots N_i$ for $i = 1, \ldots, n$

Ensure: Set of cells z, image of cells C, group number Gr, period number Pe, step number St, candidate pareto set cPs

1: $current\_group \leftarrow 1$
2: $cPs = \{\}$
3: $Gr(i) \leftarrow 0, \forall i \in N_c$
4: **for all** $pcell \in N_c$ **do**
5:      $cell \leftarrow pcell$
6:      $i \leftarrow 0$
7:      **while** $newcell = true$ **do**
8:          $x_i \leftarrow$ center point of $cell$
9:          **if** $Gr(cell) = 0$ **then**
10:             $v \leftarrow$ compute as in Eq. (9)
11:             $t \leftarrow$ compute as in Eq. (11)
12:             $p_{i+1} \leftarrow x_i + vt$
13:             $ncell \leftarrow$ cell where p_{i+1} is located
14:             $C(cell) \leftarrow ncell$
15:             $cell \leftarrow ncell$
16:             $i \leftarrow i + 1$
17:          **end if**
18:          **if** $Gr(cell) > 0$ **then**
19:             $Gr(C^j(pcell)) \leftarrow Gr(cell), j \leftarrow 0, \cdots, i$
20:             $Pe(C^j(pcell)) \leftarrow Pe(cell), j \leftarrow 0, \cdots, i$
21:             $St(C^j(pcell)) \leftarrow St(cell) + i - j, j \leftarrow 0, \cdots, i$
22:             $cell \leftarrow C(cell)$
23:             $newcell \leftarrow false$
24:          **end if**
25:          **if** $Gr(cell) = -1$ **then**
26:             $current\_group \leftarrow current\_group + 1$
27:             $Gr(C^k(pcell)) \leftarrow current\_group, k \leftarrow 0, \cdots, i$
28:             $j \leftarrow i^{th}$ value when period appears
29:             $Pe(C^k(pcell)) \leftarrow i - j, k \leftarrow 0, \cdots, i$
30:             $St(C^k(pcell)) \leftarrow j - k, k \leftarrow 0, \cdots, j - 1$
31:             $St(C^k(pcell)) \leftarrow 0, k \leftarrow j, \cdots, i$
32:             $cPs \leftarrow cPs \cup cell_k, k \leftarrow j, \cdots, i$
33:             $cell \leftarrow C(cell)$
34:             $newcell \leftarrow false$
35:          **end if**
36:      **end while**
37: **end for**

4.3   Computing the Set of Approximate Solutions

Once the run of the SCM is performed, also the approximate solutions can be detected by analyzing the given data. For instance, if an approximation of $P_{Q,\varepsilon}$ is desired, one can use the archiving technique $ArchiveUpdateP_{Q,\varepsilon}$ presented in [23, 24]. In order to prevent that all points have to be considered in the archive, one can proceed as follows:

Once the group number of the current periodic motion is discovered, the idea is to update the archive first with the periodic group of the current periodic motion and to continue with the rest of the periodic motion. Once it finds a cell which is not in $P_{Q,\varepsilon}$ it stops the procedure. This can be done since subsequent points are dominated by the current candidate solution and thus also not a member of $P_{Q,\varepsilon}$. In the worst case, however, this algorithm visits all the cells and the archive has to be updated by all candidate solutions. This is the case if the entire domain Q is equal to P_{Q_ε}. Typically, this is not the case and the number of center points considered by the archive is much lower than the total number of cells.

4.4   Error Estimates

Since SCM evaluates the entire discretized search space in one run of the algorithm, we are able to provide estimations on the maximal error that can occur in the approximation of the set of interest. Since we are particularly interested in errors of the Pareto front (i.e., errors in objective space) the following estimates are based on Lipschitz continuity.

Assume in the following that the objective function F is Lipschitz continuous on each cell, i.e.,

$$\|F(x) - F(y)\| \le L_{B(c,r)} \|x - y\|, \quad \forall x, y \in B(c, r), \tag{12}$$

where

$$B(c, r) := \{y \in \mathbb{R}^n : |c_i - y_i| \le r_i, \ i = 1, \ldots, n\} \tag{13}$$

is a cell (or generalized box) with center c and radius r, and $L_{B(c,r)}$ is the Lipschitz constant of F within $B(c, r)$. Since SCM evaluates cells at the center c and since the maximal distance on the right hand side of (12) is given for vertices of the cell, e.g., $y = c + r$, we can estimate (12) at least for unconstrained problems by

$$\|F(c) - F(y)\| \le L_{B(c,r)} \|r\|, \quad \forall y \in B(c, r). \tag{14}$$

The above formula might already be used to measure errors in image space. In the context of multi-objective optimization, however, a potential trouble is that some objectives may be in completely different ranges (e.g., the model SSW that will be

considered in the next section. See Fig. 4 for an approximation of the Pareto front).
We suggest hence to consider the error bounds for each objective, that is

$$\|F_i(c) - F_i(y)\| \le L^{(i)}_{B(c,r)}\|r\|, \quad \forall y \in B(c,r), \ i = 1, \dots, k, \tag{15}$$

where $L^{(i)}_{B(c,r)}$ is the Lipschitz constant for objective f_i. If the boxes are small enough,
one may approximate this value by the absolute value of the gradient at the center
point leading to the estimate

$$E(B(c,r),f_i) := |\nabla f_i(c)|\|r\|, \quad \forall y \in B(c,r), \ i = 1, \dots, k, \tag{16}$$

which we use in this study. As errors for the entire approximation we thus define

$$E_i := \max_{E(c,r) \in Q} E(B(c,r),f_i), \quad i = 1, \dots, k. \tag{17}$$

It remains to measure the approximation quality obtained via SCM on a particular
problem after the algorithm has been performed. For this, we think it makes sense
to measure the distance of the Pareto front to the image $F(A)$ of the archive A
of candidate solutions (e.g., the nondominated solutions) obtained by SCM. The
Inverted Generational Distance (IGD, see [41]) is widely used as a performance
indicator in multi-objective optimization, and measures the (averaged) distance of
the Pareto front to $F(A)$ as

$$IGD_p(F(A), PF) = \left(\frac{1}{M} \sum_{j=1}^{M} dist(F_j, F(A))^p \right)^{1/p}, \tag{18}$$

where $PF = \{F_1, \dots, F_M\}$ is a discretization of the Pareto front, $F(A) = \{y_1, \dots, y_N\}$, $dist(a, B) = \min_{b \in B} \|b - a\|$ the distance from point a to set B, and $p \in \mathbb{N}$. To
obtain an error bound for the objective space of each objective function f_i, one can
modify IGD as follows:

$$IGD_p^{(i)}(F(A), PF) = \left(\frac{1}{M} \sum_{j=1}^{M} dist(F_{j,i}, F(A)_i)^p \right)^{1/p}, \quad i = 1, \dots, k, \tag{19}$$

where $F_{j,i}$ denotes the i-th entry of F_j, and $F(A)_i = \{y_{1,i}, \dots, y_{N,i}\}$. Note that a finite
value of p in (19) averages the distances from F_j to $F(A)$. If this is not wanted, one
can choose $p = \infty$ leading to

$$IGD_\infty^{(i)}(F(A), PF) = \max_{i=1,\dots,M} dist(F_{j,i}, F(A)_i), \quad i = 1, \dots, k. \tag{20}$$

In this study, we will consider $p = 1$.

5   Numerical Results

In the following we present some results obtained by the cell mapping techniques and make some comparisons to other techniques.

First we consider the capability of the SCM to compute approximations of the global Pareto set. Figure 5 shows some results obtained by SCM on four MOPs (see Table 1 for the description of the problems). CONV2 is a convex bi-objective problem. The Pareto set is a curve connecting the points $(-1, -1)^T$ and $(1, 1)^T$. SSW is a bi-objective problem whose Pareto set falls into four connected components. Due to symmetries of the model two of these components (the two outer curves on the boundary of the domain) map to the same region in the Pareto front. TANAKA is a bi-objective inequality constrained problem whose Pareto front is disconnected and has convex and concave parts. Finally, CONV3 is a convex tri-objective problem. We have used a $1,000 \times 1,000$ grid to perform the cell mapping for the problems with $n = 2$ and a $100 \times 100 \times 100$ grid for CONV3. Figure 4 shows the results. In all cases SCM is able to obtain a fine grained approximation of both Pareto set and front. Table 2 shows the error estimate discussed in the previous section on six MOPs including the four ones considered in Fig. 5 on different grid sizes together with their IGD values. The match of both values (for each objective value) is almost perfect for WITTING while the IGD values for the other unconstrained problems are (much) better than the E_i values since these describe the worst case scenario. An exception is the TANAKA problem where the IGD values are worse. The reason is that this problem is constrained, and in this case the estimation made in (14) does not have to hold: it may happen that a cell contains a part of the Pareto set, but its center point is not feasible and will thus be discarded by SCM. Thus, the IGD values can get larger than the estimation made in (14) which is the case for TANAKA.

Tables 3 and 4 show comparisons of the SCM with NSGA-II [42] and MOEA/D [43], two state-of-the-art MOEAs. For the comparison we have used a budget of 60,000 function evaluations for all algorithms, and to measure the approximation quality we have used the averaged Hausdorff distance Δ_1 [44] which is more common for the comparison of different algorithms (in particular, since it gives *one* value for each problem). As anticipated, SCM cannot outperform the MOEAs (which can get similar approximation qualities even for much smaller budgets of function evaluations). Nevertheless, SCM is competing in the approximation of the Pareto *set*.

Now, we present a comparison of the SCM with an enumeration algorithm that generates all solutions with a certain precision and then we keep the nondominated solutions. The comparisons were made on CONV2 and RUDOLPH. In order to do this comparison, we use a refinement method on SCM. This can be done due to the fact that SCM returns a set of boxes and then we can use this set and apply subdivision techniques.

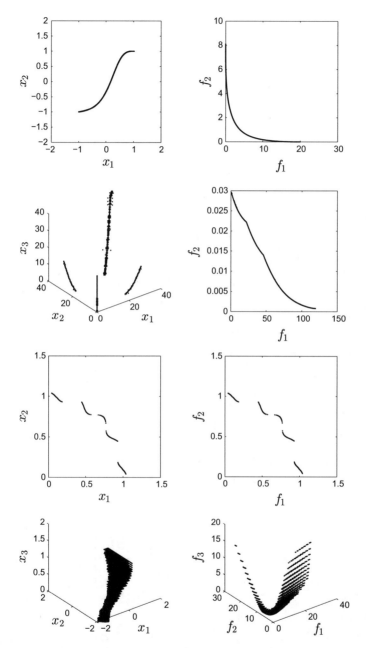

Fig. 5 Pareto sets (*left*) and fronts (*right*) obtained by the multi-objective SCM on the problems CONV2, SSW, TANAKA, and CONV3 (from *above* to *below*)

Table 2 Error of a given cell measured on maximum error (k_i err), error on the optimal points (k_i err opt) and IGD_i

Problem	Size of grid	k1 err	k2 err	k3 err	k1 err opt	k2 err opt	k3 err opt	IGD^1	IGD^2	IGD^3
WITTING [45]	100 × 100	0.223	0.223		0.1999	0.1999		0.1984	0.1984	
	200 × 200	0.1115	0.1115		0.0999	0.0999		0.0991	0.0991	
	500 × 500	0.0446	0.0446		0.04	0.04		0.0396	0.0396	
	1000 × 1000	0.0223	0.0223		0.02	0.02				
CONV2 [46]	100 × 100	10.624	0.4764		1.348	0.2424		0.8288	0.1683	
	200 × 200	5.3724	0.2391		0.6892	0.124		0.4075	0.0485	
	500 × 500	2.1635	0.0959		0.2745	0.0492		0.1081	0.0725	
	1000 × 1000	1.0842	0.048		0.1366	0.0241				
CONV3 [46]	20 × 20 × 20	105.4206	105.4206	148.9384	32.6077	14.6249	14.4099	2.1294	3.548	1.981
	30 × 30 × 30	76.124	76.124	107.5589	24.4209	8.3636	14.8069	2.0873	3.3505	2.0557
	60 × 60 × 60	41.1418	41.1418	58.1362	10.8723	4.873	6.4499	1.5386	1.5386	1.3699
	100 × 100 × 100	25.4513	25.4513	35.9657	6.5238	3.4837	3.4519			
RUDOLPH [47]	100 × 100	1.8627	1.8627		0.2843	0.2843		0.1264	0.1264	
	200 × 200	0.9413	0.9413		0.1345	0.1345		0.0494	0.0494	
	500 × 500	0.3789	0.3789		0.0566	0.0566		0.0216	0.0216	
	1000 × 1000	0.1899	0.1899		0.028	0.028				
SSW [48]	20 × 20 × 20	6	0.0031		6	0.0031		2.4	0.0005	
	30 × 30 × 30	4	0.0021		4	0.002		1.4	0.0004	
	60 × 60 × 60	2	0.0011		2	0.001		0.4	0.0002	
	100 × 100 × 100	1.2	0.0007		1.2	0.0006				
TANAKA [49]	100 × 100	0.0222	0.0222		0.0222	0.0222		0.055	0.055	
	200 × 200	0.0111	0.0111		0.0111	0.0111		0.044	0.044	
	500 × 500	0.0044	0.0044		0.0044	0.0044		0.011	0.011	
	1000 × 1000	0.0022	0.0022		0.0022	0.0022				

Table 3 Δ_1 values for the distances of the images of the candidate sets to P_Q

MOP	SCM	NSGA-II	MOEA/D
WITTING	**0.1421**	0.2925	3.9561
CONV2	0.1329	**0.1297**	0.1677
CONV3	**0.2814**	1.2607	1.7306
RUDOLPH	0.1414	**0.0232**	0.4970
SSW	**2.6533**	3.7666	32.8963

Table 4 Δ_1 values for the distances of the images of the candidate sets to $F(P_Q)$

MOP	SCM	NSGA-II	MOEA/D
WITTING	**0.1984**	0.2779	5.5826
CONV2	0.8288	**0.1578**	2.5493
CONV3	**5.5185**	16.3426	27.1944
RUDOLPH	0.1649	**0.0282**	0.7863
SSW	2.4000	**0.8969**	54.0516

For CONV2 problem, we used a 20 × 20 grid, and 3 subdivision steps using a 2 × 2 grid of test points to evaluate each cell for SCM, which leads to 7,416 function evaluations. In the case of the grid search, we used a 92 × 92 grid leading to 8,464 function evaluations. The results on terms of Δ_1 for parameter space and objective space are as follows: SCM, 0.1289 and 0.5666; grid search, 0.1529 and 1.3074 respectively. Since for Δ_1 smaller numbers represent better approximations, we can say that SCM outperforms the grid search in this example.

For RUDOLPH problem, we used a 20 × 20 grid, and 3 subdivision steps using a 3 × 3 grid of test points to evaluate each cell in the case of the SCM, which leads to 4,128 function evaluations. In the case of the grid search, we used a 65 × 65 grid leading to 4,225 function evaluations. The results in terms of Δ_1 for parameter space and objective space are as follows: SCM, 0.0414 and 0.0724; grid search, 5.9615 and 0.1246, respectively.

Figure 6 shows the results of the SCM method with subdivision techniques and Fig. 7 shows the results of the enumeration algorithm. In this case SCM shows a better performance in both problems, which is underlined by the indicator values. Also the SCM has advantages if more than 'just' the solution set is sought as the following examples demonstrate.

In some applications, it is desired to have a technique capable of computing both global and local Pareto solutions. This can be useful in cases where the global criteria does no account for all decision makers expectations [50–52].

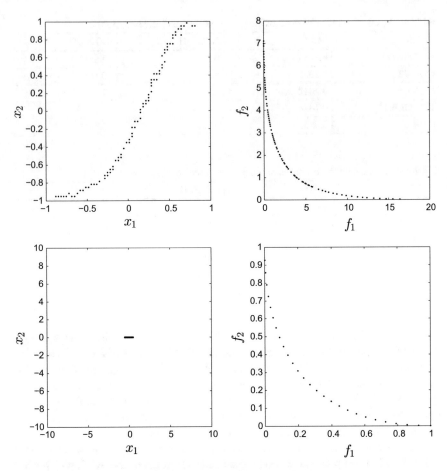

Fig. 6 Pareto sets (*left*) and fronts (*right*) using SCM with subdivision. *Above*, we can see the results of CONV2 and *below* the ones of RUDOLPH

One interesting example in this context is MOP RUDOLPH (see Table 1). In its original version proposed in [47] the Pareto set consists of nine different connected components, each of them mapping to the same Pareto front. We have modified this problem here slightly so that the Pareto set consists of one of these connected components whereas the other eight components map to a slightly higher value (more precisely, the objective values are shifted by a multiple of 0.01). Since this change in objective space is just slight, all nine components are hence potentially interesting for the decision maker if he/she is willing to accept this deterioration. Figure 8 shows the result of the SCM on this problem on a 1,000 × 1,000 grid. Note that the algorithm is capable of detecting all nine connected components, and that each component is approximated with the same quality.

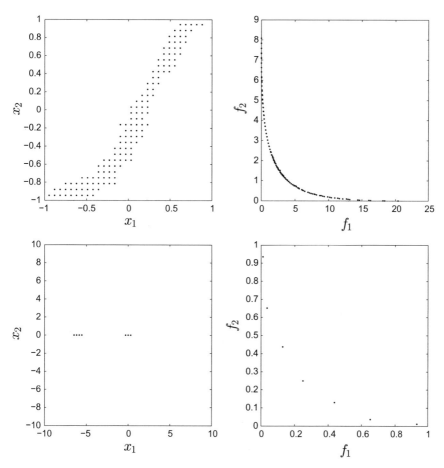

Fig. 7 Pareto sets (*left*) and fronts (*right*) using grid search on the problems CONV2 (*above*) and RUDOLPH (*below*)

Following this idea of solving multimodal problems, we now compare the SCM method with a simple multi start algorithm that uses the same descent direction than SCM. We used a 20 × 20 grid, a subdivision of 3 × 3 for each optimal cell and with 3 levels of subdivision in the case of the SCM, which leads to 12,120 function evaluations. For the multi start algorithm, we used 130 starting points leading to 12,785 function evaluations. Figure 9 shows the results of the SCM and the multi start algorithm on MOP RUDOLPH with a budget of 12,500 function evaluations. The results show that SCM is able to compute evenly spread solutions while the result of the multi start approach reveals some gaps in the fronts. The results in terms of Δ_1 for parameter space and objective space are as follows: SCM, 0.1775 and 0.3362; multi start, 0.4112 and 0.5631, respectively.

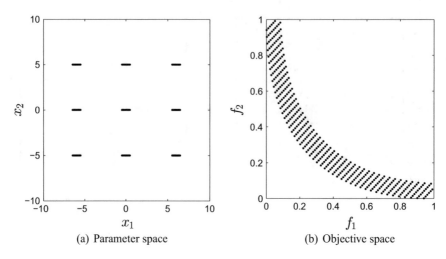

Fig. 8 Numerical result of the SCM for MOP RUDOLPH on a 1,000 × 1,000 grid

Next we address the problem to approximate the set of nearly optimal solutions. Figure 10 shows the result of the SCM on four MOPs resulting from a consideration of a 1,000 × 1,000 grid. As mentioned above, the investigation of the entire set of approximate solutions is scarce. So far, archiving techniques exist that aim for the approximation of $P_{Q,\varepsilon}$ [24], but efficient algorithms for their computations are still missing. We stress that so far many heuristics exist that utilize the concept of ε-dominance (e.g., [53–57]), however, all of them use this concept to obtain a finite size approximation of the Pareto front and *not* to obtain the set of approximate solutions. That is these works use it as a mean to improve diversity and thus get better a better approximation of the Pareto front.

In order to obtain a comparison, we have coupled NSGA-II and MOEA/D with the archiver *ArchiveUpdateP$_{Q,\varepsilon}$* [24]. The coupling can thus be viewed as an algorithm for the computation of $P_{Q,\varepsilon}$, but since both MOEAs are elitist algorithms their search naturally focuses on P_Q and not on the nearly optimal solutions. Further improvements of the evolutionary strategies can thus be obtained via further modifications of the selection operators which are, however, neither straightforward nor in the scope of this chapter. Figure 11 shows numerical results obtained by the NSGA-II variant on the same MOPs, and Tables 5 and 6 show averaged Δ_1 values of the approximation qualities in parameter and objective space, respectively. For the latter we have chosen a budget of 10,000 function calls for each algorithm. As it can be seen, SCM offers the best performance in particular for the approximation of the set of interest in decision space (which is of great interest for the decision maker as motivated in Sect. 1).

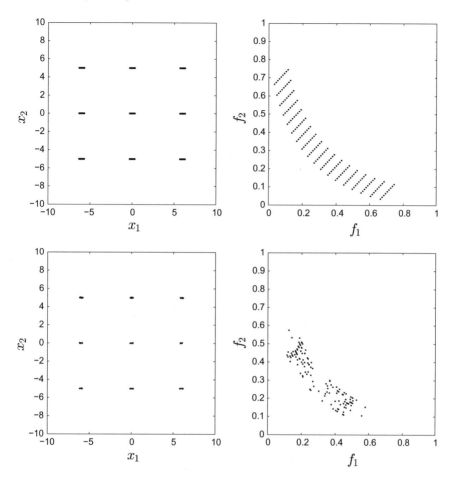

Fig. 9 Comparison of the SCM method (*above*) and multi start (*below*) on RUDOLPH

We would like to stress that $P_{Q,\varepsilon}$ is one way to define approximate solutions, but there exist other ways depending on the given situation, and that the data obtained by SCM is sufficient to comply with all of them. Figure 12 shows the approximate solutions obtained by SCM for different sets of approximate solutions. Next to $P_{Q,\varepsilon}$ we have selected the notion of Tanaka [58] and Bonnel [59].

As a hypothetical decision making problem we reconsider MOP RUDOLPH. Assume for this purpose that the DM is interested in the performance $Z = [0.17, 0.37]^T$ (measured in objective space) and further that he/she is willing to accept a deterioration of $\varepsilon = [0.1, 0.1]$. Then, for instance the representatives of the cells whose images are within the target regions can be presented to the DM leading here to 23 candidate solutions (i.e., the 'optimal' one plus another 22 nearly optimal ones) that are shown in Fig. 13. The solutions are well-spread and come in

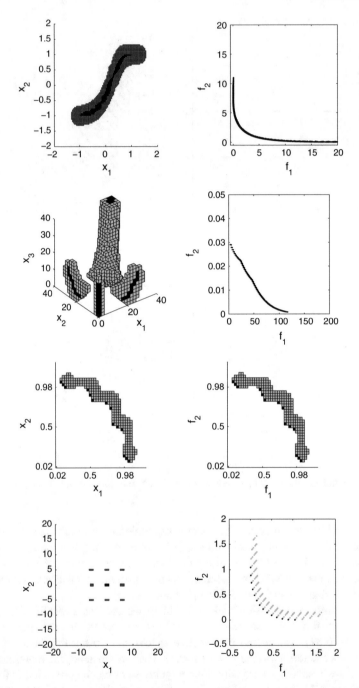

Fig. 10 Approximations of $P_{Q,\varepsilon}$ (*left*) and $F(P_{Q,\varepsilon})$ (*right*) obtained by the multi-objective SCM on the problems CONV2, SSW, TANAKA, and RUDOLPH (from *above* to *below*). In *black* the cells that contain Pareto optimal solutions, and in *green* the nearly optimal ones that do not contain a part of the Pareto set/front

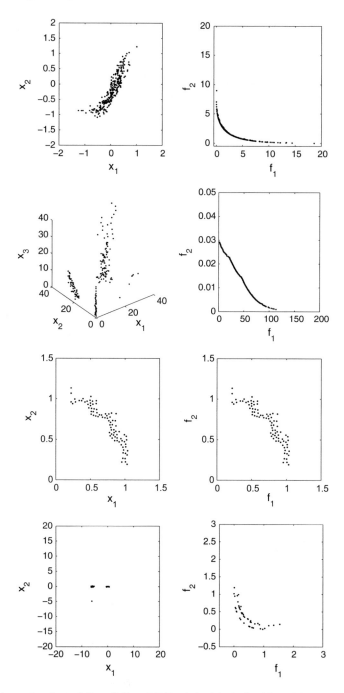

Fig. 11 Approximations of $P_{Q,\varepsilon}$ (*left*) and $F(P_{Q,\varepsilon})$ (*right*) obtained by NSGA-II coupled with *ArchiveUpdateP$_{Q,\varepsilon}$*. The problems are the same as in Fig. 10

Table 5 Δ_1 values for the distances of the candidate solution set to $P_{Q,\varepsilon}$, the best solutions in boldface

MOP	MOEA/D	NSGA-II	SCM
CONV2	0.5141	0.4628	**0.0849**
RUDOLPH	7.2438	8.0552	**0.2102**
SSW	10.8365	10.9384	**0.8660**
TANAKA	0.1462	0.1371	**0.0248**

Table 6 Δ_1 values for the distances of the images of the candidate sets to $F(P_{Q,\varepsilon})$, the best solutions in boldface

MOP	MOEA/D	NSGA-II	SCM
CONV2	7.8902	8.0027	**2.4250**
RUDOLPH	0.5090	0.7390	**0.2186**
SSW	5.8152	2.6852	**1.5000**
TANAKA	0.1462	0.1371	**0.0248**

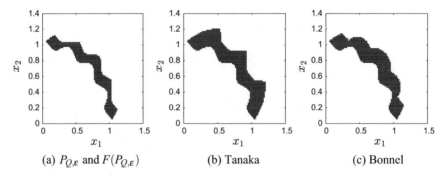

(a) $P_{Q,\varepsilon}$ and $F(P_{Q,\varepsilon})$ (b) Tanaka   (c) Bonnel

Fig. 12 Different sets of approximate solutions

Fig. 13 Hypothetical decision making problem. The figure shows the 23 boxes that could be of interest for the DM for the given setting

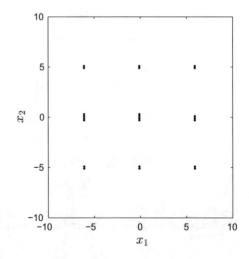

this case from all nine components of $P_{Q,\varepsilon}$. Since these components are located in different regions of the parameter space, the DM is hence given a large variety for the realization of his/her project.

6 Conclusions and Future Work

In this chapter, we have investigated cell mapping techniques for the numerical treatment of multi-objective optimization problems. Cell mapping techniques have been designed for the global analysis of dynamical systems and replace the common point-to-point by a cell-to-cell mapping via a discretization of both space and time. We have adapted the cell mapping techniques to the given context via considering dynamical systems derived from descent methods and have argued that the resulting algorithm is in particular beneficial for the thorough investigation of small problems. That is, the new algorithm is capable of detecting the global Pareto set in one run of the algorithm which is of course important for the related decision making process. For the latter, however, also other points are of potential interest such as locally optimal solutions and approximate solutions which can serve as backup solutions for the DM in case he/she is willing to accept a certain deterioration measured in objective space. The cell mapping techniques are capable of delivering all the sets after the same run of the algorithm in the same approximation quality as the computed Pareto set. While satisfactory algorithms for the computation of the Pareto set exist, such as specialized evolutionary algorithms, this does not hold for the local and approximate solutions. The cell mapping techniques presented in this work offer hence a surplus in the design of small dimensional problems by providing a thorough analysis of the problem at hand.

Though the results presented in this work are very promising, there are some points that have to be addressed in order to make the algorithm applicable to a broader class of problems. First of all, the main drawback of the cell mapping techniques is that they are restricted to small dimensional problems since the number of cells grows exponentially with the number of dimensions. Note, however, that the algorithm is highly parallelizable since the core of the algorithm is the mapping of each cell which can be realized with small effort. We expect thus that the use of massive parallelism realized e.g., via GPUs will lead to an applicability to higher dimensional problems. Further, the constraint handling techniques may be improved so that also equality constrained problems can be treated adequately. Another interesting path of future research would be to use SCM to detect a possible bias in the descent method toward the set of interest or, if possible, to design a bias free method. This could be done by considering the volumes of the basins of attractions similar as done in [60] for general dynamical systems. Bias free methods are highly wanted for memetic strategies where the aim is to get an approximation of the entire solution set. Finally, it is expected that the change from simple cell mapping techniques as used in this chapter to generalized cell mapping will offer more information to thoroughly analyze the given model.

References

1. Coverstone-Caroll, V., Hartmann, J.W., Mason, W.M.: Optimal multi-objective low-thrust spacecraft trajectories. Comput. Methods Appl. Mech. Eng. **186**(2–4), 387–402 (2000)
2. Schütze, O., Vasile, M., Coello Coello, C.A.: Computing the set of epsilon-efficient solutions in multiobjective space mission design. J. Aerosp. Comput. Inf. Commun. **8**, 53–70 (2011)
3. Das, I., Dennis, J.: Normal-boundary intersection: a new method for generating the Pareto surface in nonlinear multicriteria optimization problems. SIAM J. Optim. **8**, 631–657 (1998)
4. Eichfelder, G.: Adaptive Scalarization Methods in Multiobjective Optimization. Springer, Berlin Heidelberg (2008). ISBN 978-3-540-79157-7
5. Fliege, J.: Gap-free computation of Pareto-points by quadratic scalarizations. Math. Methods Op. Res. **59**, 69–89 (2004)
6. Hillermeier, C.: *Nonlinear Multiobjective Optimization - A Generalized Homotopy Approach*. Birkhäuser, Boston (2001)
7. Miettinen, K.: Nonlinear Multiobjective Optimization. Kluwer Academic Publishers, Boston, Massachusetts (1999)
8. Beume, N., Naujoks, B., Emmerich, M.: SMS-EMOA: multiobjective selection based on dominated hypervolume. Eur. J. Op. Res. **181**(3), 1653–1669 (2007)
9. Coello Coello, C., Lamont, G., Van Veldhuizen, D.: Evolutionary Algorithms for Solving Multi-Objective Problems, 2nd edn. Springer, Berlin (2007)
10. Deb, K.: Multi-Objective Optimization Using Evolutionary Algorithms. Wiley, Hoboken (2001)
11. Dellnitz, M., Schütze, O., Hestermeyer, T.: Covering Pareto sets by multilevel subdivision techniques. J. Optim. Theory Appl. **124**, 113–155 (2005)
12. Jahn, J.: Multiobjective search algorithm with subdivision technique. Computational Optimization and Applications **35**(2), 161–175 (2006)
13. Schütze, O., Vasile, M., Junge, O., Dellnitz, M., Izzo, D.: Designing optimal low thrust gravity assist trajectories using space pruning and a multi-objective approach. Eng. Optim. **41**(2), 155–181 (2009)
14. Gomez, M.., Martinez-Marie, T., Sanchez, S., Meziat, D.: Optimal control for wheeled mobile vehicles based on cell mapping techniques. In: 2008 IEEE Intelligent Vehicles Symposium, pp. 1009–1014 (2008)
15. Hernández, C., Naranjani, Y., Sardahi, Y., Liang, W., Schütze, O., Sun, J.Q.: Simple cell mapping method for multi-objective optimal feedback control design. Int. J. Dyn. Control **1**(3), 231–238 (2013)
16. Xiong, F.R., Qin, Z.C., Xue, Y., Schütze, O., Sun, J.Q., Ding, Q.: Multi-objective optimal design of feedback controls for dynamical systems with hybrid simple cell mapping algorithm. Commun. Nonlinear Sci. Numer. Simul. **19**(5), 1465–1473 (2014)
17. Zufiria, P.J., Martínez-Marín, T.: Improved optimal control methods based upon the adjoining cell mapping technique. J. Optim. Theory Appl. **118**(3), 657–680 (2003)
18. Hsu, C.S.: A theory of cell-to-cell mapping dynamical systems. Journal of Applied Mechanics **47**, 931–939 (1980)
19. Mersmann, O., Bischl, B., Trautmann, H., Preuss, M., Weihs, C., Rudolph, G.: Exploratory landscape analysis. In: GECCO, pp. 829–836 (2011)
20. Loridan, P.: ε-solutions in vector minimization problems. Journal of Optimization, Theory and Application **42**, 265–276 (1984)
21. Hernández, C., Sun, J.Q., Schütze, O.: Computing the set of approximate solutions of a multi-objective optimization problem by means of cell mapping techniques. In: Emmerich M. et al. (ed.) EVOLVE – A Bridge between Probability, Set Oriented Numerics and Evolutionary Computation IV, pp. 171–188. Springer (2013)
22. Pareto, V.: Cours d'Économie Politique". Lausanne, Rouge (1896)
23. Schütze, O., Coello Coello, C.A., Talbi, E.-G.: Approximating the ε-efficient set of an MOP with stochastic search algorithms. In: Gelbukh, A., Kuri Morales, A.F. (eds.), In: Mexican International Conference on Artificial Intelligence (MICAI-2007), pp. 128–138. Springer, Berlin Heidelberg (2007)

24. Schütze, O., Coello Coello, C.A., Tantar, E., Talbi, E.-G.: Computing finite size representations of the set of approximate solutions of an MOP with stochastic search algorithms. In: GECCO 2008: Proceedings of the 10th annual conference on Genetic and evolutionary computation, pp. 713–720. ACM, New York, NY, USA (2008)
25. Hsu, C.S.: Cell-to-cell mapping: a method of global analysis for nonlinear systems. Springer-Verlag, Applied mathematical sciences (1987)
26. Gu, K., Tongue, B.H.: Interpolated cell mapping of dynamical systems. J. Appl. Mech. **55**(2), 461–466 (1988)
27. Guttalu, R.S., Zufiria, P.J.: The adjoining cell mapping and its recursive unraveling, part ii: application to selected problems. Nonlinear Dyn. **4**(4), 309–336 (1993)
28. Zufiria, P.J., Guttalu, R.S.: The adjoining cell mapping and its recursive unraveling, part i: description of adaptive and recursive algorithms. Nonlinear Dyn. **4**(3), 207–226 (1993)
29. Martínez-Marín, T., Zufiria, P.J.: Optimal control of non-linear systems through hybrid cell-mapping/artificial-neural-networks techniques. Int. J. Adap. Control Signal Process. 13(4):307–319 (1999)
30. Bursal, F.H., Hsu, C.S.: Application of a cell-mapping method to optimal control problems. Int. J. Control **49**(5), 1505–1522 (1989)
31. Hsu, C.S.: A discrete method of optimal control based upon the cell state space concept. Journal of Optimization Theory and Applications **46**(4), 547–569 (1985)
32. Crespo, L.G., Sun, J.Q.: Stochastic Optimal Control of Nonlinear Dynamic Systems via Bellman's Principle and Cell Mapping. Automatica **39**(12), 2109–2114 (2003)
33. Flashner, H., Burns, T.F.: Spacecraft momentum unloading: the cell mapping approach. J. Guidance, Control Dyn. **13**, 89–98 (1990)
34. Zhu, W.H., Leu, M.C.: In: Planning Optimal Robot Trajectories by Cell Mapping, pp. 1730–1735 (1990)
35. Wang, F.Y., Lever, P.J.A.: A cell mapping method for general optimum trajectory planning of multiple robotic arms. Robot. Auton. Syst. **12**, 15–27 (1994)
36. Yen, J.Y.: Computer disk file track accessing controller design based upon cell to cell mapping (1994)
37. Bosman, P.A.N.: On gradients and hybrid evolutionary algorithms for real-valued multiobjective optimization. IEEE Trans. Evolut. Comput. **16**(1), 51–69 (2012)
38. Fliege, J., Svaiter, B.F.: Steepest descent methods for multicriteria optimization. Math. Methods Op. Res. **51**(3), 479–494 (2000)
39. Lara, A.: Using Gradient Based Information to build Hybrid Multi-objective Evolutionary Algorithms. Ph.D thesis, CINVESTAV-IPN (2012)
40. Lara, A., Alvarado, S., Salomon, S., AviGAd, G., Coello Coello, C.A., Schütze, O.: The gradient free directed search method as local search within multi-objective evolutionary algorithms. In: EVOLVE - A Bridge between Probability, Set Oriented Numerics, and Evolutionary Computation (EVOLVE II), pp. 153–168 (2013)
41. Coello Coello, C.A., Cruz Cortés, N.: Solving multiobjec tive optimization problems using an artificial immune system. Genet. Prog. Evol. Mach. **6**(2), 163–190 (2005)
42. Deb, K., Pratap, A., AGArwal, S., Meyarivan, T.: A fast and elitist multiobjective genetic algorithm: NSGA-II. IEEE Trans. Evolut. Comput. **6**(2), 182–197 (2002)
43. Zhang, Q., Li, H.: MOEA/D: a multi-objective evolutionary algorithm based on decomposition. IEEE Trans. Evolut. Comput. **11**(6), 712–731 (2007)
44. Schütze, O., Esquivel, X., Lara, A., Coello, C.A.: Using the averaged Hausdorff distance as a performance measure in evolutionary multi-objective optimization. IEEE Trans. Evolut. Comput. **16**(4), 504–522 (2012)
45. Witting, K.: Numerical Algorithms for the Treatment of Parametric Multiobjective Optimization Problems and Applications. Ph. D. thesis, University Paderborn (2012)
46. Schütze, O.: Set Oriented Methods for Global Optimization. PhD thesis, University of Paderborn (2004). http://ubdata.uni-paderborn.de/ediss/17/2004/schuetze/
47. Rudolph, G., Naujoks, B., Preuss, M.: Capabilities of EMOA to detect and preserve equivalent Pareto subsets. In: EMO'03: Proceedings of the Evolutionary Multi-Criterion Optimization Conference, pp. 36–50 (2006)

48. Schäffler, S., Schultz, R., Weinzierl, K.: A stochastic method for the solution of unconstrained vector opimization problems. Journal of Optimization, Theory and Application **114**(1), 209–222 (2002)
49. Tanaka, M.: Ga-based decision support system for multi-criteria, optimization. In: International Conference on Systems, Man and Cybernetics-2, pp. 1556–1561 (1995)
50. Krmicek, V., Sebag, M.: Functional brain imaging with multi-objective multi-modal evolutionary optimization. In: Runarsson, T.P., Beyer, H.-G., Burke, E., Merelo-Guervós, J.J., Whitley, L.D., Yao, X. (eds.) Parallel Problem Solving from Nature - PPSN IX, volume 4193 of Lecture Notes in Computer Science, pp. 382–391. Springer, Berlin, Heidelberg (2006)
51. Sebag, M., Tarrisson, N., Teytaud, O., Lefevre, J., Baillet, S., Salpétriè re, L.P., Paris, F.: Multiobjective multi-modal optimization approach for mining stable spatio-temporal patterns. In: *IJCAI* (2005)
52. Tarrisson, N., Sebag, M., Teytaud, O., Lefevre, J., Baillet, S.: Multi-objective multi-modal optimization for mining spatio-temporal patterns. In: Denis, F. (ed.) CAP, pp. 217–230. PUG (2005)
53. Deb, K., Mohan, M., Mishra, S.: Evaluating the epsilon-domination based multi-objective evolutionary algorithm for a quick computation of Pareto-optimal solutions. Evolut. Comput. **13**(4), 501–525 (2005)
54. Laumanns, M., Thiele, L., Deb, K., Zitzler, E.: Combining convergence and diversity in evolutionary multiobjective optimization. Evolutionary Computation **10**(3), 263–282 (2002)
55. Schütze, O., Laumanns, M., Coello Coello, C.A., Dellnitz, M., Talbi, E.: Convergence of stochastic search algorithms to finite size Pareto set approximations. Global Optim. **41**(4), 559–577 (2008)
56. Schütze, O., Laumanns, M., Tantar, E., Coello Coello, C.A., Talbi, E.-G.: Computing gap free Pareto front approximations with stochastic search algorithms. Evolut. Comput. **18**(1), 65–96 (2010)
57. Sierra, M., Coello Coello, C.A.: Improving PSO-based multi-objective optimization using crowding, mutation and ε-dominance. In: Proceedings of the Third International Conference on Evolutionary Multi-Criterion Optimization, pp. 505–519 (2005)
58. Tanaka, T.: A new approach to approximation of solutions in vector optimization problems. In: Fushini, M., Tone, K. (eds.) Proceedings of APORS 1994, pp. 497–504 (1995)
59. Bolintineanu, S.: (H.Bonnel). Vector variational principles; ε-efficiency and scalar stationarity. J. Convex Anal. **8**, 71–85 (2001)
60. Castillo, A., Zufiria, P.J.: Cell mapping techniques for tuning dynamical systems. In: Sun, J.Q., Luo, A.C.J. (eds.), Global Analysis of Nonlinear Dynamics, pp. 31–50. Springer (2012)

Percentile via Polynomial Chaos Expansion: Bridging Robust Optimization with Reliability

Mariapia Marchi, Enrico Rigoni, Rosario Russo and Alberto Clarich

Abstract We revise a method recently introduced by the authors for the estimation of robustness and reliability in design optimization problems with uncertainties in the input variable space. Percentile values of system output properties are estimated by means of polynomial chaos expansions used as stochastic response surfaces. The percentiles can be used as objectives or constraints in multiobjective optimization problems. We clarify the theoretical background and motivations of our approach, and we show benchmark results, as well as applications of multiobjective optimization problems solved with evolutionary algorithms. The advantages of the method are also presented.

1 Introduction

The treatment of uncertainties is a very important task in engineering design optimization. In fact, in most real-world application fields of optimization problems, design variables and problem parameters are affected by uncertainties arising from different sources such as, for instance, variations in material properties or loading conditions, measurement or manufacturing precision, or even modeling assumptions. The following questions naturally arise: What is the impact of this uncertainty on the outcome of an optimization process? Is the *best* solution found with a *deterministic* approach still the best (in terms of *reliability* and *robustness*)? By *reliability*, we mean the probability that a certain design will *not* fail to meet a predefined criterion

M. Marchi (✉) · E. Rigoni · R. Russo · A. Clarich
ESTECO S.p.A., AREA Science Park, loc. Padriciano 99, 34149 Trieste, Italy
e-mail: marchi@esteco.com

E. Rigoni
e-mail: rigoni@esteco.com

R. Russo
e-mail: russo@esteco.com

A. Clarich
e-mail: clarich@esteco.com

© Springer International Publishing AG 2017
M. Emmerich et al. (eds.), *EVOLVE – A Bridge Between Probability,*
Set Oriented Numerics and Evolutionary Computation VII,
Studies in Computational Intelligence 662, DOI 10.1007/978-3-319-49325-1_3

or performance function, called *limit state function* (LSF). Since an optimal solution often lies at the boundary of the *feasible* region, determined by the problem *constraints*, if uncertainties affect design parameters, there will be a certain probability of the optimum found to violate one or more constraints, thus compromising its reliability. By *robustness*, we mean the stability of optimization outcomes against input parameter variations, i.e., robust solutions are not very sensitive to input statistical fluctuations.

Different types of uncertainties can be considered, such as probabilistic or epistemic (see, e.g., Refs. [1, 2]). We shall focus on the former. Probabilistic uncertain input parameters are modeled by random input variables following certain probability density functions (PDFs), which represent the probability that a certain event occurs. Because of the input stochasticity, the system response is also stochastic, but its PDF is not known *a priori*. Since objectives and constraints in optimization problems are often defined in terms of output variables, the uncertainty quantification (UQ), which aims at assessing the statistical properties of random variables as precisely as possible, acquires a fundamental role. Distribution moments, such as mean and variance, can be estimated with many techniques, e.g., Monte Carlo (MC) or Latin Hypercube Sampling (LHS) [3], or the more efficient Polynomial Chaos [4] (PC). The latter is accurate while usually requiring a smaller number of function evaluations than sampling techniques. This is a crucial advantage, since computational time is one of the major bottlenecks in common design optimization processes.

References [1, 2] provide two interesting surveys about computational optimization under uncertainties and the concepts of reliability and robustness. Though intimately connected, too often these concepts are considered separately in optimization disciplines. There are two major classes of non-deterministic methods for handling uncertainties in engineering design optimization: reliability-based design optimization (RBDO) (e.g., Refs. [5–8]) and robust design optimization (RDO) (e.g., Refs. [9–12]). The deterministic optimization problem is modified, by introducing reliability indexes or robustness measures as objectives or constraints. Basic RBDO techniques seek to reduce the failure probability of a certain goal by reducing the PDF area that lies outside the feasible region boundaries, or equivalently by shifting the mean value away from constraint limits (the shape of the probability density function is assumed to remain invariant in this shift). To reach this target, probabilistic or chance constraints are added to the optimization problem. On the other hand, RDO usually aims at optimizing the mean performance while minimizing its variance. This way, the optimization problem naturally becomes multiobjective and can be directly solved by means of evolutionary and genetic algorithms, which exploit mechanisms inspired by biological evolution, such as selection, crossover, and mutation to find a set of optimal solutions in an iterative evolution process, starting from an initial set of candidate solutions (initial population) (see, e.g., [13] and references therein).

In some cases, it is not only necessary to know expectation values or variances, but also the full PDF, or its tails, in order to fully tackle the design optimization problem. *Cumulative distribution functions* (CDFs) or *percentile* values should be accurately estimated. For instance, distribution tails matter in reliability analysis, while in crash safety analyses, a prescribed percentage of designs should satisfy particular constraints due to safety norms or standards. The authors have recently developed a robust design optimization approach, which goes beyond the first two moments of output response probability density functions and can also be used to assess reliability of optimization results [14–16]. Given a function of stochastic input variables, our method aims at determining its cumulative distribution function (CDF) and percentile values by means of polynomial chaos expansion techniques. The PCE is used as a response surface model (RSM) to evaluate random samples in order to determine the CDF and the desired percentiles. Accurate estimates can be obtained with very few calls to computationally expensive simulation software. The so-determined percentiles can be used as objectives or constraints in optimization problems. Reference [14] introduces the technique and tests it on the RBDO of a single-objective problem and on a multiobjective optimization with both robustness and reliability measures. Reference [15] presents benchmarks on exact test functions and the multiobjective robust and reliability-based optimization of an H-beam under a stochastic vertical load. Reference [16] focuses on reliability aspects and shows further benchmarks before applying the method to a sizing optimization problem in structural engineering. This chapter reviews the methodology and further extends the examples of Ref. [15].

Stochastic RSMs are used to assess reliability in Geotechnics [17]. We have applied them to the context of robust multiobjective design optimization. A somewhat similar approach is described in Ref. [18], which also reviews multiobjective optimization techniques under uncertainty. PCE is also employed in stochastic finite element analysis to approximate the LSF for reliability calculations or to estimate the full probabilistic content of a system response (see, e.g., Refs. [19, 20]). By using the polynomial chaos, a key resource in RDO (as shown for instance in Ref. [21]), for the determination of properties pertinent to RBDO, a bridge is established between the two disciplines.

In Sect. 2 we summarize the basic concepts of reliability analysis and RBDO, because they are useful for understanding the following sections. We then illustrate the principles of the PCE and its application to UQ, our approach for the determination of percentile values by using the PCE as a stochastic RSM, and finally the scheme for multiobjective optimization problems. In Sect. 3, we benchmark our method by computing the PCE and percentile in the case of mathematical functions with exactly known PDFs; then, we consider its performance for the computation of failure probabilities in two classic test cases. In Sect. 4, we apply the percentile calculation to the case of multiobjective optimization problems.

2 Methods

2.1 Reliability-Based Design Optimization in Brief

The goal of reliability analysis is to find the probability of any mechanical or structural component or system of not violating certain criteria or performance functions.
Given a performance function $g(\mathbf{X})$, where $\mathbf{X} = (X_1, \ldots, X_d)$ is a tuple of d stochastic variables, the safe domain is usually defined by $g(\mathbf{X}) > 0$, while the failure
domain by $g(\mathbf{X}) \leq 0$. The boundary $g(\mathbf{X}) = 0$ is the limit state function. The system
reliability is the probability $R = P(g(\mathbf{X}) > 0)$. In most applications, its complement,
i.e., the failure probability $P_f = 1 - R = P(g(\mathbf{X}) \leq 0)$, is usually computed. P_f can
be written as the integral

$$P_f = \int_{g(\mathbf{X}) \leq 0} f_X(\mathbf{X}) \, d\mathbf{X} , \tag{1}$$

where $f_X(\mathbf{X})$ denotes the joint PDF of the stochastic variable vector.

Solving the integral in Eq. (1) is a difficult task in real-world applications. Sampling techniques, like MC, become too demanding if the probability to be found
is very small. First- or second-order reliability methods (respectively FORM and
SORM) (see, e.g., Ref. [22]) recur to various approximations. The integration domain
is simplified by transforming the original input variables to independent standard normal variables, by means of Nataf transforms [23] for instance. In the transformed
space, the PDF contours have spherical symmetry. Then, the integration function
itself is approximated. For instance, in FORM, it is linearized around the LSF point
that has the minimum distance

$$||\mathbf{u}|| = \beta \tag{2}$$

(*reliability index*) from the origin in the transformed $\mathbf{U}$-variable space ($||.||$ indicates
the norm). This yields the FORM approximation for the failure probability

$$P_{f,FORM} = \Phi(-\beta) \tag{3}$$

where $\Phi(-\beta)$ stands for the standard normal CDF. In SORM, second-order contributions are taken into account. Generalizations to the case of multiple linear state
functions are possible.

The basic reliability problem can be stated as a single-objective constrained optimization problem:

$$\begin{cases} min_\mathbf{u} \, ||\mathbf{u}|| , \\ s.t. \quad g(\mathbf{u} \leq 0) , \end{cases} \tag{4}$$

(direct reliability problem) or

$$\begin{cases} min_{\mathbf{u}} \ g(\mathbf{u}) \ , \\ s.t. \quad ||\mathbf{u}|| = \beta^r \ , \end{cases} \tag{5}$$

(inverse reliability problem).

The latter is very important for reliability-based design optimization, where probabilistic (or chance) constraints are optimized according to a given reliability threshold β^r in the so-called performance measure approach (PMA). The direct problem is instead at the basis of the reliability index approach (RIA), more expensive than the other, but capable of directly optimizing the reliability value. For a brief survey of these methods, as well as the several ways they can be integrated into an optimization process, see, e.g., Ref. [8]. Reference [8] presents the advantages of using evolutionary algorithms in RBDO, instead of classic double-loop, single-loop, or decoupled methods. In single-objective optimization problems, evolutionary algorithms are capable of reaching global reliable optima, while in multiobjective problems, a reliable approximation of the global Pareto front can be found to provide insight on the regions which are more sensitive to a desired reliability index.

2.2 Uncertainty Quantification Through Polynomial Chaos Expansions

Under specific conditions [24], a stochastic process can be expressed as a spectral expansion (*generalized polynomial chaos expansion*) based on suitable orthogonal polynomial bases, with weights associated to particular PDFs. Such expansions can be applied also if Y is a function f of a vector of d stochastic input variables $\mathbf{X} = (X_1, \ldots, X_d)$. We have then

$$Y = f(\mathbf{X}) = \sum_{i=0}^{\infty} a_i \psi_i(\mathbf{X}) \ , \tag{6}$$

with ψ_i the orthogonal polynomial basis and a_i the expansion coefficients. If Y is also dependent on deterministic variables, this dependency is accounted for by the coefficients a_i, whereas the dependency on the stochastic variables is entirely accounted for by the polynomials.

The *orthogonality condition* reads

$$\langle \psi_i \psi_j \rangle = ||\psi_i||^2 \delta_{ij} \ , \tag{7}$$

with δ_{ij} the Kronecker symbol and $||.||^2$ the squared norm associated to a scalar product

Table 1 Wiener–Askey scheme

Distribution	Weight function	Orthogonal polynomials	Support
Gaussian	$e^{-x^2/2}$	Hermite	$(-\infty, \infty)$
Uniform	1	Legendre	$(-1, 1)$
Exponential	e^{-x}	Laguerre	$(0, \infty)$
Gamma	$e^{-x}x^{\alpha}$	Generalized Laguerre	$(0, \infty)$
Beta	$(x-a)^{\alpha}(b-x)^{\beta}$	Jacobi	(a, b)

$$\langle g(\mathbf{X})h(\mathbf{X}) \rangle = \int g(\mathbf{X})h(\mathbf{X})w(\mathbf{X})d\mathbf{X} . \tag{8}$$

According to the *Wiener–Askey scheme* [25], polynomials in each specific set are orthogonal w.r.t. the weighting functions $w(\mathbf{X})$, which are proportional to certain PDFs (see Table 1). It was proven [24] that by choosing a PCE with weights corresponding to the input variable distributions, the expansion convergence rate is optimal (exponential).

For independent input variables, the PCE reduces to a tensorial product of one-dimensional orthogonal polynomials.

In computational applications, the PCE is *truncated* to a finite *chaos order* or polynomial degree k, i.e.,

$$Y = f(\mathbf{X}) \simeq \sum_{i=0}^{k} a_i \psi_i(\mathbf{X}) . \tag{9}$$

Thanks to the orthogonality condition, the mean and variance of Y are, respectively, given by

$$\mu_Y = a_0, \tag{10}$$

$$\sigma_Y^2 = \sum_{i=1}^{k} a_i^2 ||\psi_i||^2 . \tag{11}$$

This way, the problem of UQ is shifted to *finding the PCE expansion coefficients a_i* of Eq. (9). In the literature, different methods exist (like intrusive or non-intrusive Galerkin projections, collocation methods, etc., see, e.g., Ref. [26]). We determine them via a regression procedure, as in Ref. [12], by minimizing the differences between the PCE predictions (for given chaos order k) on N sampling points and the sample real output values. The sample can be arbitrarily chosen, except for a required *minimum number of points*

$$N \geq N_{min} = \frac{(k+d)!}{k!d!} \tag{12}$$

(d is the stochastic input variable space dimension) necessary to fully determine the a_i.

The accuracy of the PCE estimates of the statistical moments scales as $\exp(-N)$, where N is the number of sampling points. This is a great advantage with respect to pure sampling techniques, such as MC or LHS, where the accuracy scales, respectively, as $1/\sqrt{N}$ and $1/N$.

2.3 Percentile Calculations

Once the PCE coefficients have been determined, we can use the truncated polynomial chaos expansion of Eq. (9) as a stochastic response surface to approximate the function Y in order to get its CDF and determine percentile values. The CDF is obtained by evaluating (and sorting) the PCE responses on a LHS set of size N_{perc}. This size depends on the accuracy required for the CDF as we show in Sect. 3.2.

With respect to a pure sampling approach, where the CDF is determined by computing the true output function on the sample points, this is in principle much faster. In fact, the use of the PCE allows escaping calls to solvers, which might be very demanding from a computational point of view especially in real-world applications.

2.4 Approach for Multiobjective Optimization

The methods illustrated in Sects. 2.2 and 2.3 have been implemented within the modeFRONTIER [27] multidisciplinary and multiobjective optimization software. The uncertainty quantification flow is *nested into an optimization process*. At each optimization step, we can compute the PCE and determine the statistical properties (mean, variance, percentile values), which can be used as objectives or constraints in the main optimization flow.

The scheme proposed is shown in Fig. 1. In the UQ flow, two samples are generated: one for the PCE coefficient determination and one for the CDF and percentile computation. The first one has the smallest possible size N as the PCE coefficient determination requires *real function evaluations*, which are usually computationally demanding. However, we recommend using $N > N_{min}$ (strictly greater) to avoid overfitting problems. The second sample, as already mentioned in Sect. 2.3, has a bigger size N_{perc} and is used for non-expensive *virtual evaluations* via PCE. The last two steps of the UQ flow of Fig. 1 can be skipped if only mean and variance are needed.

The procedure illustrated does not depend on the optimization algorithm, which can be chosen on the basis of the optimization problem and other requirements.

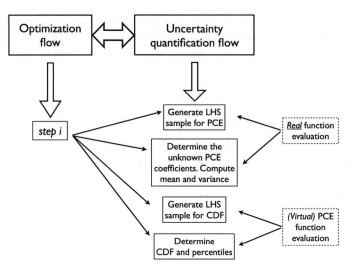

Fig. 1 Scheme of percentile determination with a polynomial chaos expansion (*uncertainty quan-tification flow*) nested into an optimization algorithm (*optimization flow*)

3 Benchmarks

We have performed a few benchmarks on analytic mathematical functions to deter-mine the accuracy of the percentile calculations for various parameter settings. We have considered several chaos orders $k = 2, 3, 4, 5$ and percentile sample sizes $N_{perc} = 100, 1\,000, 2\,000, 5\,000, 10\,000, 100\,000$. Given that in Ref. [12], it was shown that accurate estimates for means and standard deviations can already be obtained with PCEs calculated on small samples, we have considered $N = 2 \cdot N_{min}$ points in the following examples.

In order to quantify the errors on the PCE estimate of the true output values and on percentile values, we have repeated the last two steps of the UQ flow of Fig. 1 $N_{rep} = 100$ times. To be specific, for each PCE order k and each N_{perc}, we have generated 100 samples and computed the root-mean-square relative errors of the PCE output function $\Delta_{rel}(f)$ and of the computed percentile value $\Delta_{rel}(perc)$, respectively, given by:

$$\Delta_{rel}(f) = \sqrt{\frac{1}{N_{rep} \cdot N_{perc}} \sum_{i=1}^{N_{rep}} \sum_{j=1}^{N_{perc}} \left(\frac{y(\mathbf{x}_j) - \tilde{y}_i(\mathbf{x}_j)}{y(\mathbf{x}_j)} \right)^2}, \quad (13)$$

$$\Delta_{rel}(perc) = \sqrt{\frac{1}{N_{rep}} \sum_{i=1}^{N_{rep}} \left(\frac{y_{exact}^p - \tilde{y}^p}{y_{exact}^p} \right)^2}, \quad (14)$$

where $y(\mathbf{x}_j)$ denotes the value of the real output function y relative to the input variable vector $\mathbf{x}_j$, $\tilde{y}(\mathbf{x}_j)$ stands for the PCE evaluation of the vector $\mathbf{x}_j$, while y^p_{exact} and $\tilde{y}^p$ represent, respectively, the true pth percentile value and its PCE approximation on the LHS sample. $\Delta_{rel}(f)$ is an indicator of the accuracy of the PCE as stochastic response surface model of the true output function, while $\Delta_{rel}(perc)$ indicates the accuracy of the percentile estimate.

In the tests shown below, we have considered the 95th percentile, but the conclusions drawn do not depend on this particular choice.

Other indicators of the PCE accuracy (as an uncertainty quantification method) are the absolute value of the relative difference between the exact values for the output mean μ and standard deviation σ, and the PCE estimates, i.e., $\delta_{\mu,rel}$ and $\delta_{\sigma,rel}$, respectively, given by

$$\delta_{\mu,rel} = \left| \frac{\mu_{PCE} - \mu_{exact}}{\mu_{exact}} \right|, \tag{15}$$

$$\delta_{\sigma,rel} = \left| \frac{\sigma_{PCE} - \sigma_{exact}}{\sigma_{exact}} \right|. \tag{16}$$

We have also analyzed the average times necessary to generate the LHS percentile sample ($\bar{t}_{sample}$), evaluate outputs with the PCE ($\bar{t}_{eval}$), and sort the outputs in order to build the CDF and extract the percentile info ($\bar{t}_{sort}$).

3.1 Exponential Function of a Standard Normal Input Variable

We have considered the exponential function $y = e^x$ of a standard normal input variable x, with zero mean and unit standard deviation. The output function is lognormally distributed; thus, its statistical properties are known exactly. The function y has been expanded in a basis of orthogonal Hermite polynomials.

In Table 2, we show the PCE accuracy indicators $\delta_{\mu,rel}$ and $\delta_{\sigma,rel}$ of Eqs. (15)–(16) and the root-mean-square relative errors $\Delta_{rel}(f)$, $\Delta_{rel}(perc)$ of Eqs. (13)–(14) for the various k, N, and N_{perc} considered. Four trends can be observed: *Trend one*: every indicator, except for $\Delta_{rel}(f)$ (which we will discuss in the second trend), assumes smaller values for increasing polynomial chaos degree k. This fact is reasonable, since an improved quality of the PCE approximation of an exponential function is expected for bigger k values. *Trend two*: $\Delta_{rel}(f)$ has an oscillating behavior for increasing k. This could be due to the difficulty of approximating an exponential function with a polynomial expansion. *Trend three*: $\Delta_{rel}(f)$ is approximately constant (apparently it oscillates around a value) for increasing sizes N_{perc}. This also sounds reasonable. Indeed, $\Delta_{rel}(f)$ indicates the quality of the PCE as a surrogate model of the real output function, and we do not expect the PCE prediction accuracy to be dependent on N_{perc}, unless some points belong to regions where the approximation

Table 2 *Exponential function of a standard normal input variable.* From left to right: polynomial chaos order k, number of PCE sample points N, PCE accuracy indicators $\delta_{\mu,rel}$ and $\delta_{\sigma,rel}$ of Eqs. (15)–(16), percentile sample size N_{perc}, and root-mean-square relative errors $\Delta_{rel}(f)$ and $\Delta_{rel}(perc)$ of Eqs. (13)–(14)

k	N	$\delta_{\mu,rel}$	$\delta_{\sigma,rel}$	N_{perc}	$\Delta_{rel}(f)$	$\Delta_{rel}(perc)$
2	6	0.156	0.387	100	0.773	0.22666
				1000	1.144	0.22722
				2000	1.231	0.22730
				5000	1.238	0.22727
				10000	1.231	0.22727
				100000	1.368	0.22726
3	8	0.042	0.178	100	1.72	0.03514
				1000	3.40	0.03482
				2000	3.20	0.03490
				5000	3.30	0.03491
				10000	3.77	0.03494
				100000	3.90	0.03494
4	10	0.018	0.092	100	1.45	0.01912
				1000	1.70	0.00723
				2000	1.40	0.00698
				5000	2.13	0.00696
				10000	1.56	0.00697
				100000	2.10	0.00697
5	12	0.0067	0.0216	100	2.206	0.02159
				1000	2.814	0.00274
				2000	3.275	0.00191
				5000	4.060	0.00167
				10000	4.197	0.00167
				100000	4.192	0.00164

is particularly poor. *Trend four*: $\Delta_{rel}(perc)$ has approximately constant values for increasing N_{perc} values for $k = 2$, 3, while they decrease until saturation for $k = 4$, 5. This could be due to a competition between the PCE approximation accuracy as a RSM and the improvements in the CDF for increasing percentile sample sizes N_{perc}. For small k values, the effects of the PCE accuracy prevail and no advantage is found by increasing the sample size N_{perc}. At first, for bigger k values, the percentile computation becomes more accurate (thanks to the improved CDF derived from a bigger sample). Subsequently, this effect saturates for $N_{perc} > 1\,000$.

The effects of k on the CDF accuracy are shown in Fig. 2, where the approximate CDF obtained by a PCE evaluation of $N_{perc} = 1\,000$ sample points (with $k = 2$ or $k = 3$) is compared to the true CDF. In this, as well as in other examples, we have found that at least $k = 3$ should be used, in order to get a decent approximation.

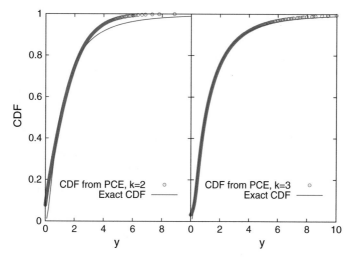

Fig. 2 CDF of the *exponential function of a standard normal input variable*: comparison of the exact CDF (*solid line*) and PCE estimates (*empty circles*) for $k = 2$ (*left panel*) and $k = 3$ (*right panel*)

Fig. 3 Average times (in seconds), $\bar{t}_{sample}$ (*solid circles*), $\bar{t}_{eval}$ (*solid squares*), and $\bar{t}_{sort}$ (*solid triangles*) versus percentile sample size N_{perc} for the $k = 3$ PCE percentile computation of the *exponential function of a standard normal variable*. Logarithmic scales have been used for both axes. Lines are included as a visual aid

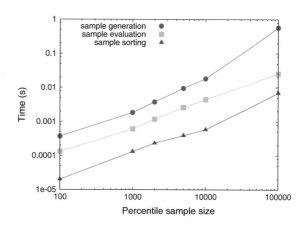

In Fig. 3, we show how the average sample generation, evaluation, and sorting times scale with the number of sampling points N_{perc}. The example provided is for the case $k = 3$. The most expensive process is the percentile sample generation. The least expensive process is the value sorting for the construction of the sample CDF. The sample evaluation time lies in between the others, and we do expect this trend to be confirmed also in real-world applications since this is the time for the *virtual* function evaluation by means of the PCE.

3.2   Ratio of Two Lognormal Input Variables

As a second test case, we have considered a two-dimensional problem ($d = 2$): the ratio of two lognormally distributed input variables. The expectation value and standard deviation of the logarithm of each input variable are 1 and 0.5 for both variables. The output function has also a lognormal distribution, corresponding to a normal variable with zero mean and $\sqrt{2}/2$ standard deviation.

Lognormal random variables can be straightforwardly *transformed*, by means of a nonlinear variable transform, into standard normal variables. Indeed, if x is a lognormal random variable with probability density function $f_X = \frac{1}{\sqrt{2\pi V}x}e^{-\frac{(\ln x - M)^2}{2V}}$, where M and V denote the expectation value and variance of the associated normal variable, then $z = \frac{\ln x - M}{\sqrt{V}}$ is a standard normal variable. This way, it is possible to apply the Wiener–Askey scheme and expand the considered function in a basis of Hermite polynomials in the transformed input variable space.

In Table 3, we report the test results. In this benchmark, all the indicators appear to assume smaller values for increasing k, although the differences found between the $k = 3$ and $k = 4$ cases are not very big. As found in Table 2, $\Delta_{rel}(f)$ has approximately constant values for given k and increasing percentile sample sizes. For $\Delta_{rel}(perc)$, at variance with the exponential function benchmark, we can observe a greater accuracy of the PCE estimates for increasing N_{perc} values, until a saturation effect (found only for $k = 2$ and $k = 3$).

In Fig. 4, we show the effects of the sample size N_{perc} on the CDF were found with our method. In the four panels, we compare the exact CDF and the approximate CDF obtained with a $k = 3$ PCE and $N_{perc} = 100, 1\,000, 10\,000, 100\,000$ (from left to right, top to bottom). As expected, we can observe an increasing smoothness and accuracy of the approximate CDF for increasing values of N_{perc}. Starting from $N_{perc} = 1\,000$, the exact and approximate CDFs almost coincide, except for small deviations in the lower range of probability.

In regard to average times for the sample generation, evaluation, and value sorting, we have found no significant difference from the case of Sect. 3.1.

3.3   Failure Probability for a Mechanical Component
and a Cantilever Beam

We have investigated two classic examples of reliability analysis, also considered in Ref. [28], with the aim of benchmarking the performances of our method for the estimation of *failure probabilities*. In the following, the two examples will be referred to as *"case 1"* and *"case 2."* In both, the stochastic input space dimension is $d = 2$ and the input variables are normally distributed.

Table 3 Results for the *ratio of two lognormal input variables*. From left to right: polynomial chaos order k, number of PCE sample points N, PCE accuracy indicators $\delta_{\mu,rel}$ and $\delta_{\sigma,rel}$ of Eqs. (15)–(16), percentile sample size N_{perc}, and root-mean-square relative errors $\Delta_{rel}(f)$ and $\Delta_{rel}(perc)$ of Eqs. (13)–(14)

k	N	$\delta_{\mu,rel}$	$\delta_{\sigma,rel}$	N_{perc}	$\Delta_{rel}(f)$	$\Delta_{rel}(perc)$
2	12	0.003	0.101	100	0.743	0.07972
				1 000	0.749	0.03245
				2 000	0.778	0.03012
				5 000	0.901	0.02449
				10 000	0.868	0.02307
				100 000	0.810	0.02166
3	20	0.0013	0.021	100	0.531	0.09064
				1 000	0.857	0.03047
				2 000	0.697	0.02751
				5 000	0.620	0.01653
				10 000	0.617	0.01397
				100 000	0.631	0.01015
4	30	0.0011	0.014	100	0.333	0.09697
				1 000	0.607	0.03329
				2 000	0.675	0.02041
				5 000	0.491	0.01475
				10 000	0.517	0.00964
				100 000	0.469	0.00301
5	42	2.58E-4	0.003	100	0.178	0.10823
				1 000	0.077	0.03459
				2 000	0.156	0.02316
				5 000	0.163	0.01414
				10 000	0.132	0.00944
				100 000	0.125	0.00341

Case 1 concerns a mechanical component. As a limit state function $g(\mathbf{X})$, we have taken the difference between the strength $X_1 \sim N(200, 20)$ MPa and the maximum stress $X_2 \sim N(150, 10)$ MPa. $N(\mu, \sigma)$ denotes a normal distribution with mean μ and standard deviation σ. The LSF reads

$$g(\mathbf{X}) = X_1 - X_2 \ . \tag{17}$$

Case 2 deals with a cantilever beam with rectangular section. As a LSF, we have taken the difference between a maximum allowable displacement value "$D_0 = 3$" and the tip displacement, i.e.,

$$g(\mathbf{X}) = D_0 - \frac{4L^3}{Ewt}\sqrt{\left(\frac{X_2}{t^2}\right)^2 + \left(\frac{X_1}{w^2}\right)^2} \ , \tag{18}$$

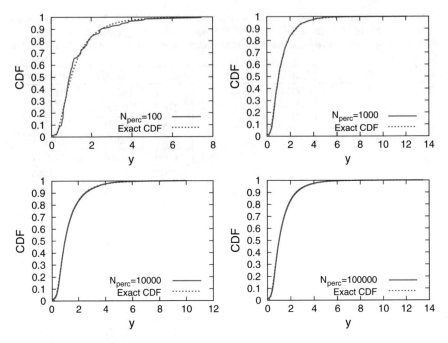

Fig. 4 CDF of the *ratio of two lognormal variables*: comparison of the exact CDF (*dashed line, blue online*) and a PCE estimate (*solid line, red online*) for $k = 3$ and different values of N_{perc} as indicated in the labels

with elasticity module $E = 30 \cdot 10^6$ psi, beam length "$L = 100$," and beam section dimensions "$w = 2$" and "$t = 4$." $X_1 \sim N(500, 100)$ lb and $X_2 \sim N(1\,000, 100)$ lb are vertical and lateral external normal forces applied to the beam.

In both cases, failure modes correspond to $g(\mathbf{X}) \leq 0$. Sample estimates of the failure probabilities P_f are determined with the formula:

$$P_f = \frac{n_f}{N_{tot}}, \tag{19}$$

where n_f corresponds to the number of failure events (i.e., the number of sample points for which $g(\mathbf{X}) \leq 0$) and N_{tot} represents the total number of events, i.e., the sample size.

We have determined the PCE expansions of the LSFs and performed tests as in Sects. 3.1 and 3.2.

Since the LSF of Eq. (17) is a linear combination of two normally distributed variables, its mean value, standard deviation, percentiles, and CDF are exactly known. The same is not true for *case 2*. Thus, in order to benchmark our results, we have also generated a *target* LHS of 1 million points and computed sample estimates of the desired quantities.

Table 4 *Maximum stress limit state function*: relative errors $\delta_{\mu,rel}$, see Eq. (15), and $\delta_{\sigma,rel}$, see Eq. (16). For comparison, target values computed on a target LHS sample of 1 million points are provided in the last row

k	N	$\delta_{\mu,rel}$	$\delta_{\sigma,rel}$
2	12	1.97E-8	4.97E-8
3	20	6.49E-10	5.39E-9
4	30	6.95E-8	9.02E-8
5	42	2.24E-8	1.70E-7
Target LHS		7.77E-9	1.91E-4

Table 4 reports mean value and standard deviation errors with respect to the exact values for *case 1*, see Eqs. (15)–(16), as well as the deviations found in the target LHS sample of 1 million points. The mean value accuracy of the PCE estimates is very high and oscillates around the value of the target LHS one, whereas the standard deviation estimate through the PCE seems to be much more accurate than the target sample estimate.

Table 5 reports the root-mean-square relative errors $\Delta_{rel}(f)$ and $\Delta_{rel}(perc)$ of Eqs. (13)–(14), as well as the failure probabilities P_f and LHS P_f. Both failure probabilities are derived from the sample size N_{perc} by using Eq. (19). P_f is obtained by evaluating the sample points with the PCE approximation, whereas LHS P_f indicates the failure probability found with the true LSF evaluation.

In this simple case (linear LSF, function of two independent normal random variables), the PCE approximates the true LSF very well, as shown by the small values obtained for $\Delta_{rel}(f)$ in Table 5. As a consequence, the error on the 95th percentile $\Delta_{rel}(perc)$ is also very small even for $k = 2$. Moreover, the failure probabilities computed with the PCE approximation (P_f) and with the true function evaluation on the same N_{perc} sample (LHS P_f) have identical values. The failure probability values must be compared with the target LHS calculation estimate $P_f = 0.01263$ and the first-order reliability method (FORM) [28] (which is *exact* for a linear LSF) value $P_f = 0.0127$. The relative error between the target LHS failure probability and the exact one is $\sim 0.6\%$, while the estimates provided by the PCE used as a stochastic response surface yields relative errors of at most $\sim 3\%$ for $k > 2$ and $N_{perc} \geq 1\,000$. The relative error becomes comparable to the target LHS as N_{perc} increases.

Since for *case 2* exact values are not known, the only comparison possible is with the target LHS outcomes. Thus, Table 6 reports directly the mean value μ and standard deviation σ estimated with the PCE. The target LHS estimates are shown in the last row. The agreement between PCE outcomes and target LHS is good.

Table 7 shows the root-mean-square relative error on the function $\Delta_{rel}(f)$, the 95th percentile value of the LSF ($f^{0.95}$), as well as the failure probabilities P_f and LHS P_f (same notation as in Table 5). Although the errors on the PCE function estimate are three to five magnitude orders greater than magnitude orders in *case 1*, the percentile values found differ from the target LHS results by only a small percent (the relative difference, being $\sim 10^{-4}$). Also the relative agreement between P_f and the LHS P_f

Table 5 *Mechanical component: maximum stress limit state function.* From left to right PCE order k, the number of PCE sample points N, percentile sample size N_{perc}, root-mean-square relative errors on the function estimate $\Delta_{rel}(f)$ and on the percentile $\Delta_{rel}(perc)$, see Eqs. (13)–(14), failure probability computed on the sample of N_{perc} points with the PCE approximation of the LSF (P_f) and with the true function evaluation (LHS P_f). For comparison, the target LHS results on a sample of 1 million points are $\Delta_{rel}(perc) = 6.32\text{E-}4$ and $P_f = 0.01263$

k	N	N_{perc}	$\Delta_{rel}(f)$	$\Delta_{rel}(perc)$	P_f	LHS P_f
2	12	100	7.713E-6	0.02782	0.0111	0.0111
		1 000	1.32E-5	0.00991	0.01179	0.01179
		2 000	1.14E-5	0.00739	0.012295	0.012295
		5 000	5.08E-5	0.00515	0.012856	0.012856
		10 000	1.90E-4	0.00363	0.012661	0.012661
		100 000	4.19E-4	0.00112	0.0126708	0.0126709
3	20	100	1.97E-6	0.03239	0.0106	0.0106
		1 000	3.67E-6	0.00973	0.01232	0.01232
		2 000	1.37E-5	0.00724	0.01277	0.01277
		5 000	8.33E-5	0.00453	0.01227	0.01227
		10 000	1.42E-5	0.00352	0.012629	0.012629
		100 000	1.00E-4	0.00108	0.012695	0.012695
4	30	100	3.17E-5	0.03063	0.0128	0.0128
		1 000	4.28E-5	0.01110	0.01281	0.01281
		2 000	1.11E-4	0.00743	0.01278	0.01278
		5 000	2.35E-4	0.00484	0.012786	0.012786
		10 000	2.06E-4	0.00392	0.012615	0.012615
		100 000	0.00138	0.00119	0.0126999	0.0127
5	42	100	1.13E-4	0.03388	0.0136	0.0136
		1 000	3.46E-5	0.01104	0.01269	0.01269
		2 000	1.11E-4	0.00842	0.01277	0.01277
		5 000	6.35E-5	0.00842	0.012934	0.012934
		10 000	4.53E-4	0.00404	0.012741	0.012741
		100 000	6.47E-4	0.00110	0.0126768	0.0126768

Table 6 *Cantilever beam limit state function:* mean value μ and standard deviation σ estimated with a PCE approximation of the LSF. For comparison, target values computed on a LHS sample of 1 million points are provided in the last row

k	N	μ	σ
2	12	0.6613652863493148	0.37416866059775356
3	20	0.6610289197443749	0.37193471743269313
4	30	0.6609597977068523	0.37200793971216195
5	42	0.6608838533734408	0.37191515245843537
Target LHS		0.660877783494823	0.37191038371188223

Table 7 *Cantilever beam: tip displacement limit state function.* From left to right PCE order k, the number of PCE sample points N, percentile sample size N_{perc}, root-mean-square relative error on the function estimate $\Delta_{rel}(f)$, see Eq. (13), 95th percentile value of the LSF ($f^{0.95}$), failure probability computed on the sample of N_{perc} points with the PCE approximation of the LSF (P_f) and with the true function evaluation (LHS P_f). For comparison, the target LHS computation on a sample of 1 million points yields $f^{0.95} = 1.2617203229641354$ and $P_f = 0.040894$

k	N	N_{perc}	$\Delta_{rel}(f)$	$f^{0.95}$	P_f	LHS P_f
2	12	100	0.11371	1.26186	0.0417	0.0411
		1 000	0.79136	1.26487	0.04143	0.04099
		2 000	9.08047	1.26566	0.04171	0.041115
		5 000	1.53963	1.26546	0.041464	0.040942
		10 000	4.54342	1.26493	0.041522	0.040974
		100 000	3.21347	1.26555	0.0415129	0.0409903
3	20	100	0.02679	1.26241	0.0406	0.0407
		1 000	0.06632	1.26085	0.04091	0.04098
		2 000	0.66060	1.26091	0.040825	0.040865
		5 000	0.07576	1.26106	0.040962	0.040984
		10 000	0.32231	1.26097	0.04105	0.041104
		100 000	0.82031	1.26106	0.0409503	0.0409975
4	30	100	0.01741	1.26081	0.04	0.0401
		1 000	0.19688	1.26167	0.04083	0.04086
		2 000	0.28078	1.26232	0.041015	0.04113
		5 000	0.06368	1.26109	0.040878	0.040968
		10 000	0.10976	1.26115	0.040828	0.040924
		100 000	1.75674	1.26136	0.040946	0.041027
5	42	100	0.00641	1.26436	0.0411	0.0411
		1 000	0.06967	1.26043	0.04144	0.0414
		2 000	0.02962	1.26211	0.04128	0.04125
		5 000	0.12767	1.26111	0.041066	0.041052
		10 000	0.09531	1.26123	0.041092	0.041078
		100 000	0.19915	1.26106	0.0410295	0.0410132

is quite good (in the worst cases, it is $\sim 10^{-2}$), and the accord with the target failure probability (0.040894) is also satisfactory. To be noted is that FORM and SORM results [28] are $P_f = 0.04054$ and $P_f = 0.04098$.

Figure 5 shows the CDF of the LSFs of *case 1* and *case 2*. In both panels, we compare the PCE results obtained with $k = 3$, $N = 20$, and $N_{perc} = 100\,000$ (empty circles) with a target CDF obtained with real function evaluations on the target LHS of 1 million points (solid lines). In the left panel, the exact PDF of the LSF is also shown (dashed line). The agreement between the PCE prediction and the exact CDF is good, except for small deviations (emphasized by the use of a logarithmic scale (basis 10) on the CDF axis) occurring at small probability values. The agreement

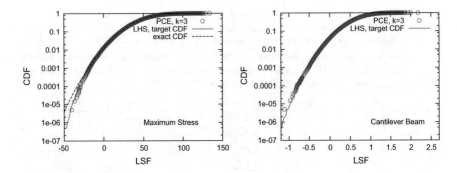

Fig. 5 CDF of the limit state function for the maximum stress of a mechanical component (*left panel*) and a cantilever beam tip displacement (*right panel*). PCE estimates (obtained with $k = 3$, $N = 20$, and $N_{perc} = 100\,000$), denoted by *empty circles*, are compared to LHS target results (*solid line*) and, for the mechanical component case, also to the exact CDF value (*dashed line, left panel*)

between the target CDF obtained with true function evaluations and the PCE ($k = 3$) approximation obtained with only 20 real function evaluations is good in both cases.

4 Multiobjective Optimization of an H-Beam

In this section, we show an example of *multiobjective RDO*, in which percentiles are used as objectives and constraints during the optimization. When reliability aspects are considered in multiobjective optimization problems, the reliable front may be different from the deterministic Pareto front. In particular, by requiring more reliable solutions, the front is expected to move further inside the feasible objective space, as shown in Ref. [8] by comparing outcomes of a deterministic run on the "CON-STR" problem of Ref. [29] with RBDO results obtained by introducing probabilistic constraints and requiring a given reliability to be respected. The same trends were observed in Ref. [16] by performing a RDO on the same problem and considering percentile values of the constraints, with a probability threshold corresponding to the desired reliability. The complementary issue of the robustness of the Pareto front in RDO problems is addressed for instance in Ref. [21], where the sensitivity of solutions w.r.t. input perturbations is investigated.

For this chapter, we have considered an H-beam with *three input variables*, i.e., the H section dimensions (Web thickness $a \in [1, 10]$ mm, flange width $b \in [50, 150]$ mm, and flange depth $c \in [50, 250]$ mm, see left panel of Fig. 6), and *one input parameter*, i.e., the external load $F = 2\,000$ N (vertically applied in the middle of the beam, see right panel of Fig. 6). The Young's modulus $E = 2 \cdot 10^5$ MPa, the beam density $\rho = 8\,000$ Kg/m^3, and the beam length $L = 3$ m are *constant parameters*.

The *output variables* of the problem are the maximum stress

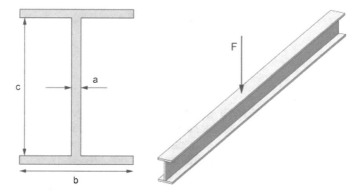

Fig. 6 *H-beam*: section (*left panel*) and load (*right panel*)

$$\sigma = \frac{FL}{4I}\left(a + \frac{c}{2}\right), \tag{20}$$

the beam deflection

$$\delta = \frac{FL^3}{48EI}, \tag{21}$$

and the beam weight

$$W = (2ab + ca)L\rho, \tag{22}$$

where

$$I = \frac{ac^3}{12} + 2\left[\frac{ba^3}{12} + ab\left(\frac{a}{2} + \frac{c}{2}\right)^2\right] \tag{23}$$

represents the beam section moment of inertia of the plane area.

We have considered a two-objective optimization problem (minimization of the weight and of the maximum deflection) subject to two constraints (a maximum allowed weight and a maximum allowed stress). The *deterministic optimization problem* reads:

$$\begin{cases} min \ \{W, \delta\} \\ s.t. \ \ W \le 20 \, \text{Kg}, \\ \quad \quad \sigma \le 100 \, \text{MPa}. \end{cases} \tag{24}$$

We have also considered a stochastic optimization problem, where we have imputed all the uncertainty to be due to the applied load and the variable a representing the Web thickness of the beam. We have assumed both F and a to be *normally distributed*. For F, we have taken a standard deviation $\sigma_F = 100$N, while for a, we have taken a constant standard deviation corresponding to $\sim 1\%$ of the central value of the range of values allowed for *a*. The *stochastic optimization prob-lem* reads:

$$\begin{cases} min\ \{W\ ,\ \delta^{99.999}\} \\ s.t.\quad W \leq 20\,\text{Kg}\ , \\ \qquad \sigma^{99.999} \leq 100\,\text{MPa}\ , \end{cases} \tag{25}$$

where $\delta^{99.999}$ and $\sigma^{99.999}$, respectively, denote the 99.999th percentile value of the maximum beam deflection δ and the maximum stress σ. Taking the 99.999th per-centile of the maximum deflection δ instead of δ's mean value enhances the mini-mization condition. In fact, only one in a hundred thousand designs will *fail* to have a deflection value smaller than the allowed 99.999th percentile deflection value.

For the multiobjective optimization, we have used the genetic algorithm NSGA-II [29] with an initial population of 50 random individuals (created by a random DOE node) for 60 generations, with crossover and mutation probabilities equal to 0.9 and 0.05, respectively, a distribution index of 2, and an automatic scaling for mutation probability. First, we have performed the deterministic run. Then, we have performed a robust design optimization. For the PCE, we have used $k = 3$ and $N = 20$ and we have chosen $N_{perc} = 100\,000$ for the percentile computation. In Fig. 7, we show a modeFRONTIER workflow for the optimization problem of Eq. (25).

In Fig. 8, we compare the deterministic (circles) and the stochastic Pareto fronts (squares) (weight values are plotted on the *y*-axis, while deflection values are plotted on the *x*-axis). As expected, the stochastic Pareto front is pushed away from the infeasible region in the objective space.

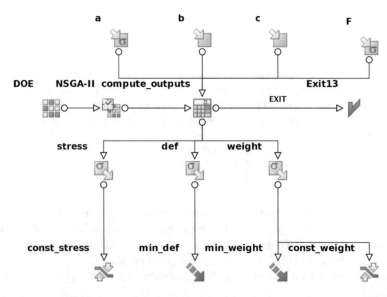

Fig. 7 *H-beam: optimization workflow.* From top to bottom: input variables (*a, b, c, F*), DOE, optimizer and calculator nodes, output variables σ, δ, W (respectively denoted by *stress, def, weight*), constraints on $\sigma^{99.999}$ and W (*const_stress, const_weight*), and objectives on $\delta^{99.999}$ and W (*min_def, min_weight*)

Fig. 8 *H-beam*: comparison of deterministic (*circles*) and stochastic (*squares*) Pareto fronts (weight versus beam deflection)

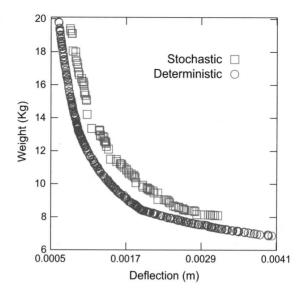

In order to compare the reliability of solutions found with the deterministic and stochastic run, we have picked up the minimum weight Pareto front design of each simulation (represented by the rightmost circle and square in Fig. 8) and we have evaluated (with the real function) a LHS sample of 2 000 designs drawn around each solution by considering the distribution of F. In Fig. 9, we compare the probability distribution functions of the deterministic and stochastic deflection (min_def). The

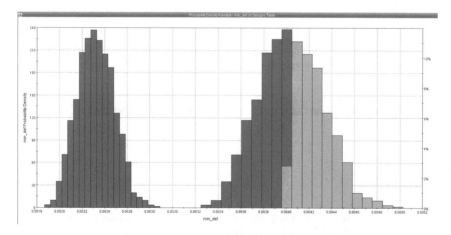

Fig. 9 *H-beam*: comparison of deterministic (*right*) and stochastic (*left*) empirical probability distribution functions sampled around the minimum weight solution of the deterministic and stochastic run of Fig. 8. The light-colored (*orange online*) histograms indicate a violation of the stress constraint

histograms representing the PDF based on the RDO optimization lie on the left, while the PDF from the deterministic optimization is on the right. We can observe that approximately one-half of the designs distributed around the deterministic solution breaks the stress constraint (as indicated by the light-colored histograms, orange online), while the probability density function sampled around the stochastic solution is pushed away from the failure region, as expected, and no design in this sample violates the stress constraint.

5 Conclusions

We have reviewed a method which combines reliability and robustness, in multiobjective design optimization, by means of polynomial chaos expansions. The PCE is used as a stochastic response surface model to estimate desired percentile values of output functions of random input variables. This technique has been benchmarked on mathematical test functions both for the computation of percentile values and for the prediction of accurate failure probabilities.

The use of the PCE as a stochastic metamodel yields a high accuracy in the evaluation of the statistical properties, while decreasing the number of real function evaluations necessary to determine the full probability content of a system response w.r.t. pure sampling approaches. This represents an important advantage in real-world applications, where computationally demanding engineering software is usually needed for real function evaluations.

The calculation of percentiles can be nested into a robust design optimization process. The overall optimization problem can be solved in terms of percentile quantities, thus guaranteeing that objectives and constraints meet predefined probability thresholds. Thus, reliability and robustness measures can be considered at the same time in a single optimization run. The method can be naturally used with multiple objectives. As an application, we have shown the multiobjective optimization of an H-beam. However, this approach is also applicable to more complex engineering systems.

Acknowledgements The authors would like to thank Cristina Belli (ESTECO S.p.A.) for the manuscript revision.

References

1. Schuëller, G.I., Jensen, H.A.: Computational methods in optimization considering uncertainties - an overview. Comput. Methods Appl. Mech. Eng. **198**, 2–13 (2008)
2. Beyer, H.-G., Sendhoff, B.: Robust optimization - a comprehensive survey. Comput. Methods Appl. Mech. Eng. **196**(33–34), 3190–3218 (2007)

3. McKay, M.D., Conover, W.J., Beckman, R.J.: A comparison of three methods for selecting values of input variables in the analysis of output from a computer code. Technometrics **21**, 239–245 (1979)
4. Wiener, N.: The homogeneous chaos. Am. J. Math. **60**, 897–936 (1938)
5. Youn, B.D., Choi, K.K.: An investigation of nonlinearity of reliability-based optimization approaches. J. Mech. Design **126**, 403–411 (2004)
6. Deb, K., Padmanabhan, D., Gupta, S., Mall, A.K.: Handling uncertainties through reliability–based optimization using evolutionary algorithms. In: Fourth International Conference on Evolutionary Multi–Criterion Optimization (EMO 2007), LNCS, vol. 4403, pp. 66–88 (2007)
7. Daum, D. A., Deb, K., Branke, J.: Reliability–based optimization for multiple constraints with evolutionary algorithms. In: 2007 IEEE Congress on Evolutionary Computation (CEC 2007), pp. 911–918 (2007)
8. Deb, K., Gupta, S., Daum, D., Branke, J., Mall, A.K., Padmanabhan, D.: Reliability-based design optimization. IEEE Trans. Evol. Comput. **13**, 1054–1074 (2009)
9. Branke, J.: Creating robust solutions by means of an evolutionary algorithm. In: Eiben, A.E., Bäck, T., Schoenhauer, M., Schwefel, H.-P. (eds.) Parallel Problem Solving from Nature, vol. 1498, pp. 119–128. Springer, New York (1998)
10. Matton, C.A., Messac, A.: Pareto frontier based concept selection under uncertainty, with visualization. Optim. Eng. **6**, 85–115 (2005)
11. Deb, K., Gupta, H.: Introducing robustness in multi-objective optimization. Evol. Comput. **14**, 463–494 (2006)
12. Poles, S., Lovison, A.: A polynomial chaos approach to multiobjective optimization. In: Dagstuhl Seminar Proceedings 09041, Hybrid and Robust Approaches to Multiobjective Optimization (2009). http://drops.dagstuhl.de/opus/volltexte/2009/2000
13. Marler, R.T., Arora, J.S.: Survey of multi-objective optimization methods for engineering. Struct. Multidisc. Optim. **26**, 369–395 (2004)
14. Clarich, A., Russo, R., Marchi, M., Rigoni, E.: Reliability–based design optimization applying polynomial chaos expansion: theory and applications. In: 10th World Congress on Structural and Multidisciplinary Optimization, Orlando, Florida, USA, 19–24 May (2013)
15. Marchi, M., Rigoni, E., Russo, R., Clarich, A.: Percentile via polynomial chaos expansion: bridging robust optimization and reliability. In: International Conference on EVOLVE 2013, A Bridge between Probability, Set Oriented Numerics, and Evolutionary Computation, Extended Abstract Proceedings, Leiden, NL, 10–13 July 2013. ISBN 978-2-87971-118-8, ISSN 2222-9434
16. Marchi, M., Rizzian, L., Rigoni, E., Russo, R., Clarich, A.: Combining robustness and reliability with polynomial chaos techniques in multiobjective optimization problems: use of percentiles. In: Cunha, A., Caetano, E., Ribeiro, P., Müller G. (eds.) Proceedings of the 9th International Conference on Structural Dynamics, EURODYN 2014, Porto, Portugal, 30 June–2 July 2014
17. Li, D., Chen, Y., Lu, W., Zhou, C.: Stochastic response surface method for reliability analysis of rock slopes involving correlated non-normal variables. Comput. Geotech. **38**, 58–68 (2011)
18. Filomeno Coelho, R., Bouillard, P.: Multi-objective reliability-based optimization with stochastic metamodels. Evol. Comput. **19**, 525–560 (2011)
19. Sudret, B., Der Kiureghian, A.: Comparison of finite element reliability methods. Probab. Eng. Mech. **17**, 337–348 (2002)
20. Sudret, B., Berveiller, M., Lemaire, M.: A stochastic finite element procedure for moment and reliability analysis. Eur. J. Comput. Mech./Rev. Européenne de Mécanique Numér. **15**, 825–866 (2006)
21. Molina-Cristobal, A., Parks, G.T., Clarkson, P.J.: Finding robust solutions to multi-objective optimisation problems using polynomial chaos. In: Proceedings of the 6th ASMO UK/ISSMO Conference on Engineering Design Optimization Oxford, UK, 3–4 July 2006
22. Ditlevsen, O., Madsen, H.O.: Structural Reliability Methods. Wiley, Chichester (1996). internet edition v. 2.3.7, June–September 2007
23. Nataf, A.: Détermination des distribution dont les marges sont données. C. R. de l'Académie des Sci. **225**, 42–43 (1962)

24. Xiu, D., Karniadakis, G.E.: The Wiener–Askey polynomial chaos for stochastic differential equations. SIAM J. Sci. Comput. **24**, 619–644 (2002)
25. Askey, R., Wilson, J.: Some basic hypergeometric polynomials that generalize Jacobi polynomials. Mem. Amer. Math. Soc. AMS **54**(319), 121–142 (1985)
26. Loeven, G.J.A., Witteveen, J.A.S., Bijl, H.: Probabilistic collocation: an efficient non–intrusive approach for arbitrarily distributed parametric uncertainties. In: 45th AIAA Areospace and Sciences Metting and Exhibit, AIAA paper 2007–317, Reno, Nevada (2007)
27. http://www.esteco.com
28. Du, X.: First order and second reliability methods. In: Probabilistic Engineering Design, Missouri S&T, ME 360 (2009)
29. Deb, K., Pratap, A., Agarwal, S., Meyarivan, T.: A fast and elitist multiobjective genetic algorithm: NSGA-II. IEEE Trans. Evol. Comput. **6**, 182–197 (2002)

Part II
Evolutionary Methods in Computational Game Theory

Evolutionary Equilibrium Detection in Multicriteria Games

Réka Nagy and D. Dumitrescu

Abstract Since most real life decisions are multiobjective, multicriteria games offer a more realistic modeling of real-life interactions. Although several equilibrium concepts have been proposed for solving multicriteria games, equilibria detection has not received much attention. Generative relations are proposed to characterize multicriteria equilibria. An evolutionary method based on generative relations is proposed for detecting various multicriteria equilibria: Nash-Pareto, Ideal Nash and Pareto equilibria. Numerical experiments on discrete and continuous games indicate the potential of the proposed approach.

1 Introduction

In standard non-cooperative games players are agents whose only goal is to maximize their own payoffs. This is an overly simplified model of reality. Real life players usually make decisions considering more than one, often conflicting, criteria. These criteria in most of the cases are not measured by the same unit, they can not be just aggregated into one single criterion.

Games with multiple criteria offer more accurate real life models. Several multicriteria equilibrium concepts have been proposed [11, 14] and vast research addressed their existence [2, 3, 15, 16] but the detection of these equilibria did not receive much attention.

Detecting the Nash equilibrium is a computationally hard problem [4]. Equilibrium detection in standard games can be viewed as a multiobjective optimization problem, where the payoff of each player is to be maximized. Since evolutionary algorithms are powerful tools for solving multiobjective optimization problems, they can also be applied for finding game equilibria [5, 6]. In an equilibrium detection

R. Nagy (✉) · D. Dumitrescu
Babeş-Bolyai University, 1 Mihail Kogălniceanu Street, Cluj Napoca, Romania
e-mail: reka@cs.ubbcluj.ro

D. Dumitrescu
e-mail: ddumitr@cs.ubbcluj.ro

© Springer International Publishing AG 2017　　　　　　　　　　　　　　83
M. Emmerich et al. (eds.), *EVOLVE – A Bridge Between Probability,*
Set Oriented Numerics and Evolutionary Computation VII,
Studies in Computational Intelligence 662, DOI 10.1007/978-3-319-49325-1_4

problem the payoff functions are the objective functions. Thus in the corresponding optimization problem the number of variables is equal to the number of objectives and are given by the number of players in the game.

Most equilibrium types can be characterized by *generative relations* [6]. These are algebraic tools that allow the comparison of two strategy profiles with respect to a certain equilibrium. Generative relations guide the search towards the certain equilibrium.

Our goal is to propose generative relations for multicriteria equilibrium types and develop an evolutionary method for detecting equilibria in multicriteria games.

2 Non-cooperative Games

Non-cooperative games, also called strategic games, model simple forms of interactions between rational players. In non-cooperative games the goal of each player is to maximize her payoff. The value of the payoff function depends on the decisions taken simultaneously by all players [9].

Definition 1 A finite non-cooperative game is defined as a system

$$\Gamma = (N, S_i, u_i, i = 1, \ldots, n),$$

where:

- N represents a set of n players, $N = \{1, \ldots, n\}$;
- for each player $i \in N$, S_i represents the set of actions available to her, $S_i = \{s_{i_1}, s_{i_2}, \ldots, s_{i_m}\}$;

$$S = S_1 \times S_2 \times \cdots \times S_n$$

  is the set of all possible situations of the game;
- an element of S is called a strategy profile;
- for each player $i \in N$, $u_i : S \to R$ represents the payoff function.

We denote by (s_{i_j}, s^*_{-i}) the strategy profile obtained from s^* by replacing the strategy of player i with s_{i_j} i.e.

$$(s_{i_j}, s^*_{-i}) = (s^*_1, s^*_2, \ldots, s^*_{i-1}, s_{i_j}, s^*_{i+1}, \ldots, s^*_n).$$

In standard Computational Game Theory the following propositions are assumed:

- Players choose their strategies simultaneously, without collaborating with each other. The profit of each player is affected by the strategies chosen by the other players as well.
- All players are rational, meaning that the objective of each player is to maximize her payoff.

- Players have common knowledge of the game and the rationality of the other players.

2.1 Nash Equilibrium

The central solution concept in Computational Game Theory is Nash equilibrium, introduced in [8]. Nash equilibrium captures a state in which individual players act according to their incentives, maximizing their own payoff. A strategy profile is a Nash equilibrium if no player has the incentive to unilaterally deviate from her strategy. Once all players are playing Nash equilibrium, it is in interest of every player to stick to her strategy [1, 7].

Definition 2 The strategy s^* is a Nash equilibrium if and only if the inequality

$$u_i(s_i, s^*_{-i}) - u_i(s^*) \leq 0, \ \forall s_i \in S_i, \forall i \in N$$

holds.

2.2 Pareto Equilibrium

The concept of Pareto equilibrium is inspired from the solution of Multiobjective Optimization problems. Pareto equilibrium consists of the Pareto-optimal outcomes of the game, and is based on the Pareto dominance relation.

2.2.1 Pareto Dominance:

Let x' and x'' be two m-dimensional real vectors.

x' weakly Pareto-dominates x'', and we write, $x' \succeq_P x''$, if:

$$x' \succeq_P x'' \Leftrightarrow x'_i \geq x''_i, \forall i \in \{1, \ldots, m\}.$$

x' Pareto-dominates the solution x'', and we write, $x' \succ_p x''$, if:

$$x' \succ_p x'' \Leftrightarrow x' \succeq_P x'' \text{ and } \exists j \in \{1, \ldots, m\} : x'_j > x''_j.$$

Definition 3 A strategy profile s Pareto dominates the strategy profile s^* if and only if

$$s \succ_P s^* \Leftrightarrow u_i(s) \succ_P u_i(s^*)$$

Remark 1 By an abuse of notation we use the same notation, $\succeq_P$ and $\succ_P$, for the weak and strong Pareto domination between strategy profiles and real vectors.

Definition 4 A strategy profile s^* is Pareto non-dominated, or Pareto-efficient, if there exists no $s \in S$ such that $s \succ_P s^*$. In other words, a strategy profile is Pareto non-dominated if no player can increase her payoff without decreasing the payoff of other players.

Definition 5 The Pareto equilibrium of the game is the set of Pareto non-dominated strategy profiles.

Thus, the Pareto-equilibrium of the game consists of the optimal outcomes. Very often this is an infinite set of solutions.

3 Multicriteria Games - Games with Vector Payoffs

Multicriteria games (or games with vector payoffs) are natural extensions of standard non-cooperative games and offer a more realistic model for real life situations.

Definition 6 A finite strategic multicriteria game is a non-cooperative game with vector payoff functions. For each player $i \in N$

$$u_i : S \to R^{r(i)}$$

represents the multicriteria payoff function, where $r(i) \in \mathbb{N}$ is the number of criteria for player i.

We consider multicriteria games where each player has the same number of criteria, i.e.

$$r_1 = \cdots = r_n = r$$

If all players have only one criterion ($r_i = 1, \forall i \in N$) then we have a standard non-cooperative game.

Any multicriteria game G with r criteria is composed of r standard non-cooperative games: $G_1, \ldots, G_r$.

4 Equilibria in Multicriteria Games

Equilibrium concepts in multicriteria games and their existence has been widely studied [2, 3, 11, 15, 16].

The first and most popular equilibrium concept proposed for multicriteria games is the *Pareto-Nash equilibrium* [11], but other solution concepts such as *Perfect equilibrium* [3] or *Ideal Nash equilibrium* [14], etc., have also been introduced.

4.1 Pareto-Nash Equilibrium

The most studied multicriteria equilibrium concept is the *Pareto-Nash equilibrium* introduced in [11]. The *Pareto-Nash equilibrium* concept is an extension of the Nash equilibrium for standard games and is based on Pareto dominance (Sect. 2.2.1).

Since the multicriteria Pareto-Nash equilibrium is based on the Pareto dominance relation, we can distinguish weak and strong Pareto-Nash equilibria.

Definition 7 A strategy profile $s^* \in S$ is a *weak Pareto-Nash equilibrium* if and only if the following condition holds

$$u_i(s^*) \succeq_P u_i((s_i, s^*_{-i})), \ \forall s_i \in S_i, \forall i \in N.$$

Definition 8 A strategy profile $s^* \in S$ is a *strong Pareto-Nash equilibrium* if and only if the following condition holds

$$u_i(s^*) \succ_P u_i((s_i, s^*_{-i})), \ \forall s_i \in S_i, \forall i \in N.$$

Remark 2 If not stated otherwise, by multicriteria Pareto-Nash equilibrium we refer to the strong Pareto-Nash equilibrium.

4.2 Ideal Nash Equilibrium

The *Ideal Nash equilibrium* is introduced in [14] and is also studied in [10]. Multicriteria games are often viewed as strategic interactions between organizations. Each player i corresponds to an organization with r members and each criterion corresponds to the concerns of a different member of the organization.

The concept of Ideal Nash equilibrium captures the following reasoning: a choice of strategy of the organization i is supposed to be taken by common agreement of all the r members with the objective to maximize the payoff for each member of the organization. The goal of each member is to maximize its own profit. Also, the payoffs of the members depend on the strategy choices of other organizations as well.

The Ideal Nash equilibrium of a multicriteria game G consists of those solutions that are Nash equilibria in the single-criterion games, that constitute the multicriteria game G.

Definition 9 Let $s^* \in S$ be a strategy profile of a multicriteria game. A strategy profile s^* is an ideal Nash equilibrium if and only if the following condition holds:

$$u_i^j(s^*) > u_i^j((s_i, s^*_{-i})) \forall s_i \in S_i, \forall i \in N, \forall j \in 1, \ldots r.$$

The idea of viewing players as organizations is realistic, since in many real life situations decisions are influenced by several individuals with different objectives.

4.3 *Multicriteria Pareto Equilibrium*

The Pareto equilibrium of a standard game consists of the Pareto-optimal solutions. The Multicriteria Pareto Equilibrium is a generalization of this concept.

Definition 10 A strategy profile s^* is a multicriteria Pareto equilibrium if for all players the outcomes generated by a strategy profile Pareto-dominate the outcomes generated by any other strategy profile. More formally:

$$\forall s \in S, \forall i = 1, \ldots, n : \ u_i(s^*) \succ_P u_i(s)$$

holds.

5 Evolutionary Equilibrium Detection in Multicriteria Games

Vast research addresses the existence of equilibria in multicriteria games, but the detection of equilibria has not received much attention. Various game equilibria may be characterized by generative relations on the set of game strategies [6]. The idea is that the non-dominated strategies with respect to the generative relation equal (or approximate) the equilibrium set.

5.1 *Generative Relations*

Let us consider a relation $\mathcal{R}$ over $S \times S$.
A strategy x is non dominated with respect to relation $\mathcal{R}$ if

$$\nexists y \in S : (x, y) \in \mathcal{R}.$$

Let us denote by *NDR* the set of non-dominated strategies with respect to relation $\mathcal{R}$. A subset $S' \subset S$ is non-dominated with respect to $\mathcal{R}$ if and only if

$$\forall s \in S', s \in NDR.$$

Definition 11 Relation $\mathcal{R}$ is said to be a *generative relation* of an equilibrium type if and only if the set of non-dominated strategies with respect to $\mathcal{R}$ equals the set of equilibria.

5.2 Generative Relation for Pareto-Nash Equilibrium

Let s^*, $s \in S$ be two strategy profiles. Let $k_{PN}(s^*, s)$ denote the number of players which by deviating from s^* towards s can increase a payoff for a criterion without decreasing the payoffs for the other criteria:

$$k_{PN}(s^*, s) = card\{i \in N, u_i((s_i, s^*_{-i})) \succ_P u_i(s^*), s^*_i \neq s_i\}.$$

The value $k(s^*, s)_{PN}$ is a relative quality measure of s and s^* - with respect to the multicriteria Pareto-Nash equilibrium.

Definition 12 The strategy profile s^* is better than the strategy profile s with respect to the multicriteria Pareto-Nash equilibrium, and we write $s^* \succ_{PN} s$ if the inequality

$$k_{PN}(s^*, s) < k_{PN}(s, s^*)$$

holds.

Definition 13 The strategy profile s^* is called Pareto-Nash non-dominated if there exists no strategy profile s such that

$$s \succ_{PN} s^*.$$

Proposition 1 *Let s^* be a strategy profile. s^* is a multicriteria Pareto-Nash equilibrium if and only if the quality measure $k_{PN}(s^*, s)$ is zero for any strategy profile $s \in S$. More formally:*

$$s^* \text{ is a Pareto-Nash equilibrium} \Leftrightarrow k_{PN}(s^*, s) = 0, \forall s \in S.$$

Proof Let the strategy profile $s^* \in S$ be a multicriteria Pareto-Nash equilibrium. Suppose that there exists a strategy profile $s \in S$ such that $k_{PN}(s^*, s) = w$ where $w \in 1 \dots, n$. This means that there exists a player $i, i \in N$ such that: $u_i((s_i, s^*_{-i})) \succ_P u_i(s^*)$ and $s_i \neq s^*_i$. This means that the strategy profile s^* is not a multicriteria Pareto-Nash equilibrium, which is a contradiction.

Let $s^* \in S$ be a strategy profile. Suppose that $\forall s \in S$ $k_{PN}(s^*, s) = 0$. This means that $\forall i \in N, \forall s_{i_j} \in S_i : u_i(s^*) \succ_P u_i((s_{i_j}, s^*_{-i}))$, so s^* is a multicriteria Pareto-Nash equilibrium.

Proposition 2 *All non-dominated solutions with respect to relation $\succ_{PN}$ are multicriteria Pareto-Nash equilibria.*

Proof Let $s^* \in S$ be a non-dominated solution with respect to the relation $\succ_{PN}$ and not a multicriteria Pareto-Nash equilibrium. This means that $\exists s \in S, \exists i : u_i((s_i, s^*_{-i})) \succ_P u_i(s^*)$. Let the strategy profile $q = (s_i, s^*_{-i})$. So $k(s^*, q) \geq 1$. But in the same time $k(q, s^*) = 0$. This means that $q \succ_{PN} s^*$, which is a contradiction.

Proposition 3 *All the multicriteria Pareto-Nash equilibria are non-dominated with respect to the relation* $\succ_{PN}$.

Proof Let the strategy profile $s^* \in S$ be a multicriteria Pareto-Nash equilibrium. Suppose that $\exists s : s \succ_{PN} s^*$ such that $k_{PN}(s, s^*) < k_{PN}(s^*, s)$. But since s^* is a multicriteria Pareto-Nash equilibrium, $k_{PN}(s^*, s) = 0$. So, $k_{PN}(s, s^*) < 0$, which is not possible.

Proposition 4 *Let s^* be a strategy profile. s^* is a multicriteria Pareto-Nash equilibrium if and only if it is non-dominated with respect to the relation* $\succ_{PN}$.

The relation $\prec_{PN}$ can be considered as the *generative relation of the multicriteria Pareto-Nash equilibrium*.

5.3 Generative Relation for Ideal Nash Equilibrium

Let $s^*, s \in S$ be two strategy profiles. Let $k_{iN}(s^*, s)$ denote the number of players which by deviating from s^* towards s can increase their payoff for any criterion:

$$k_{iN}(s^*, s) = card\{i \in N, \exists j \in 1, .., r : u_i^j((s_i, s_{-i}^*)) > u_i^j(s^*), s_i^* \neq s_i\}.$$

The value $k_{iN}(s^*, s)$ is a relative quality measure of s and s^* - with respect to the ideal Nash equilibrium.

Definition 14 The strategy profile s^* is better than the strategy profile s with respect to the ideal Nash equilibrium, and we write $s^* \succ_{iN} s$, if the inequality

$$k_{iN}(s^*, s) < k_{iN}(s, s^*)$$

holds.

Definition 15 The strategy profile s^* is called ideal Nash non-dominated if there exists no strategy profile s such that

$$s \succ_{iN} s^*.$$

Proposition 5 *Let s^* be a strategy profile. s^* is an ideal Nash equilibrium if and only if the quality measure $k_{iN}(s^*, s)$ is zero for any strategy profile $s \in S$. More formally:*

$$s^* \text{ is an ideal Nash equilibrium} \Leftrightarrow k_{iN}(s^*, s) = 0, \forall s \in S.$$

Proof Let the strategy profile $s^* \in S$ be an ideal Nash equilibrium. Suppose that there exists a strategy profile $s \in S$ such that $k_{iN}(s^*, s) = w$ where $w \in 1 \ldots, n$. This means that there exists a player i, $i \in N$ such that: $\exists j : u_i^j((s_i, s_{-i}^*)) \geq u_i^j(s^*)$ and

$s_i \neq s_i^*$. This means that the strategy profile s^* is not an ideal Nash equilibrium, which is a contradiction.

Let $s^* \in S$ be a strategy profile. Suppose that $\forall s \in S \ k_{iN}(s^*, s) = 0$. This means that $\forall i \in N, \forall s_{i_j}, \forall l \in S_i : u_i^l(s^*) > u_i^l((s_{i_j}, s_{-i}^*))$, so s^* is an ideal Nash equilibrium.

Proposition 6 *All non-dominated solutions with respect to relation $\succ_{iN}$ are ideal Nash equilibria.*

Proof Let $s^* \in S$ be a non-dominated solution with respect to the relation $\succ_{iN}$ and not an ideal Nash equilibrium. This means that $\exists s \in S, \exists i, j : u_i^j((s_i, s_{-i}^*)) \geq u_i^j(s^*)$. Let the strategy profile $q = (s_i, s_{-i}^*)$. So $k_{iN}(s^*, q) \geq 1$. But in the same time $k_{iN}(q, s^*) = 0$. This means that $q \succ_{iN} s^*$, which is a contradiction.

Proposition 7 *All the multicriteria ideal Nash equilibria are non-dominated with respect to the relation $\succ_{iN}$.*

Proof Let the strategy profile $s^* \in S$ be an ideal Nash equilibrium. Suppose that $\exists s :$ $s \succ_{iN} s^*$ such that $k_{iN}(s, s^*) < k_{iN}(s^*, s)$. But since s^* is an ideal Nash equilibrium, $k_{iN}(s^*, s) = 0$. So, $k_{iN}(s, s^*) < 0$, which is not possible.

Proposition 8 *Let s^* be a strategy profile. s^* is an ideal Nash equilibrium if and only if it is non-dominated with respect to the relation $\succ_{iN}$.*

The relation $\prec_{iN}$ can be considered as the *generative relation of the ideal Nash equilibrium*.

5.4 Generative Relation for Multicriteria Pareto Equilibrium

Let $s^*, s \in S$ be two strategy profiles.

Definition 16 The strategy profile s^* is better than s with respect to the multicriteria Pareto equilibrium, and we write $s^* \succ_{mP} s$ if:

$$\forall i \in 1, .., n \ u_i(s^*) \succ_P u_i(s).$$

Definition 17 The strategy profile s^* is non-dominated with respect to the relation $\succ_{mP}$, if and only if there is no strategy profile $s \in S$ such that $s \succ_{mP} s^*$.

The relation $\succ_{mP}$ can be considered as the *generative relation for multicriteria Pareto equilibrium*. In other words the non-dominated strategy profiles with respect to the relation $\succ_{mP}$ induce the multicriteria Pareto equilibrium.

5.5 Evolutionary Equilibrium Detection in Multicriteria Games

Games can be viewed as multiobjective optimization problems, where the payoffs of the participating players are to be maximized. All of the objectives to be optimized are uniform and equally important.

An appealing technique is the use of generative relations and evolutionary algorithms for detecting equilibrium strategies. The payoff of each player is treated as an objective and the generative relation induces an appropriate dominance concept, which is used for fitness assignment purpose. Evolutionary multiobjective algorithms are thus suitable tools for detecting game equilibria.

A population of strategy profiles is evolved. A chromosome is an n-dimensional vector representing a strategy profile $s \in S$. The payoff for each player i is an r dimensional vector.

The strategy profiles are compared with the help of generative relations. The non-dominated individuals from the population of strategy profiles at iteration t may be regarded as the current equilibrium approximation. Subsequent application of the search operators is guided by a specific selection operator induced by the generative relation. Successive populations produce new approximations of the equilibrium front, which hopefully are better than the previous ones. The process will finally converge to the multicriteria equilibrium induced by the generative relation.

Remark 3 For evolutionary equilibria detection any state of the art evolutionary multiobjective algorithm can be used. Our goal is to focus on the detected equilibrium types and not on the algorithm used for their detection.

5.6 Crowding Differential Algorithm

Differential Evolution (DE), proposed in [12], is an evolutionary algorithm inspired by simplex methods. DE has been proposed to solve real-parameter optimization problems on continuous domains. The advantage of DE consists in its simplicity and efficiency.

The structure of a DE algorithm is similar to the structure of a genetic algorithm. A randomly initialized population is improved using selection, mutation and crossover operations.

At each generation for each individual a so-called *trial vector* is created. The trial vector is constructed by adding the differences between randomly selected elements of the population to another element. The most common variant for creating a trial vector $v = (v_1, \ldots, v_n)$ from parent x_i is referred as strategy rand/1/bin and is given by:

$$v_j = \begin{cases} x_{r_{3_j}} + F(x_{r_{1_j}} - x_{r_{2_j}}), & \text{with probability } CR, \\ x_{i_j}, & \text{with probability } 1 - CR \end{cases}$$

where r_1, r_2 and r_3 are distinct random indices from $\{1, 2, \ldots N\}$, $F \in [0, 2]$ is a scaling factor, $CR \in [0, 1]$ is the crossover probability, and N is the size of the population.

If the trial vector has a better fitness value than the parent, the parent is replaced in the population. This way the average fitness of the population is never worsened.

The Crowding based Differential Evolution (CDE) algorithm [13] is a multiobjective algorithm that enhances the standard DE algorithm with a crowding mechanism. The only difference from the standard DE algorithm is that an offspring, instead of the parent, is compared to the most similar element from the population.

The CDE algorithm is summarized in Algorithm 1. The relation R stands for the generative relation of the desired equilibrium.

Algorithm 1 CDE for evolutionary equilibrium detection.

initialize population $\mathscr{P}$
repeat
  for all $i \in \{1, 2, \ldots, pop\_size\}$ **do**
    create candidate v
    find the element w most similar to v in the design space
    if w R v **then**
      replace v with w
    end if
  end for
until a stopping condition is met

---

6 Numerical Experiments

In our numerical experiments we consider a simple two player discrete game and a more complex continuous game. For equilibria detection we use the Crowding Differential Evolution algorithm (Sect. 5.6) with the following parameter settings: $CR = 0.3$, $F = 0.5$, $pop\_size = 100$.

6.1 Discrete Games

Let us consider a two-player two-criteria discrete game: *game D*. Each player has two strategies, S_1 and S_2 and the payoff matrix for the game is represented in Table 1.

The game D is composed of two unicriterial games D_1 and the D_2. The payoff matrices of the unicriterial games are depicted in Table 2. The Nash equilibrium for the game D_1 consists of two strategy profiles (S_1, S_1) and (S_2, S_2) with payoff values of $(10, 10)$ and $(7, 7)$ respectively. The game D_1 has only one Nash equilibrium the strategy profile (S_1, S_1) with payoff values of $(5, 5)$.

Table 1 Payoff matrix for the two player two-criteria game D

	Player2	
Player1	S_1	S_2
S_1	$[(10,5);(10,5)]$	$[(0,8);(7,0)]$
S_2	$[(7,0);(0,8)]$	$[(7,1);(7,1)]$

Table 2 The payoff matrix for the D_1 and D_2 unicriterial games

	Player2	
Player1	S_1	S_2
S_1	$(10,10)$	$(0,7)$
S_2	$(7,0)$	$(7,7)$

(a) Game D_1

	Player2	
Player1	S_1	S_2
S_1	$(5,5)$	$(8,0)$
S_2	$(0,8)$	$(1,1)$

(b) Game D_2

Table 3 The detected Pareto, Nash-Pareto and Ideal Nash equilibria for the multicriteria game D

Equilibrium	Strategies	Payoffs
Pareto	(S_1, S_1)	$[(10,5);(10,5)]$
Nash-Pareto	(S_1, S_1)	$[(10,5);(10,5)]$
	(S_2, S_2)	$[(7,1);(7,1)]$
Ideal Nash	(S_1, S_1)	$[(10,5);(10,5)]$

The detected multicriteria equilibria of the multicriteria game D are summarized in Table 3. The Pareto equilibrium of the multicriteria game is the strategy profile (S_1, S_1). This is the optimal solution of the game, the outcome consists of the most preferred payoffs for both players in both criteria. The Pareto-Nash equilibrium of game D consists of two strategy profiles: (S_1, S_1) and (S_2, S_2). Note that the Pareto-Nash equilibrium corresponds to the Nash equilibrium of the unicriterial component D_1. The Ideal Nash equilibrium by definition consists of those solutions that are Nash equilibria in all the unicriterial components of the game. In this case we have one such strategy profile: (S_1, S_1).

We can conclude that the Pareto-Nash and Ideal Nash equilibria do not always correspond. The Ideal Nash equilibrium is a refinement of the Pareto-Nash equilibrium, and in most cases provides more optimal solutions (given that the game has an Ideal Nash equilibrium).

6.2 Continuous Games

Consider a continuous two-player two-criteria game: *game G* [10]:

$$G = (\{1, 2\}, (S_i)_{i \in \{1,2\}}, (u_i)_{i \in \{1,2\}}),$$

where

$$S_1 = (-1, 1),$$
$$S_2 = [0, 1],$$
$$u_1(x, y) = (-x^2 + y^2, y \cos(\tfrac{\pi}{2}x)),$$
$$u_2(x, y) = (u_{21}(x, y), u_{22}(x, y)), \text{ where}$$

$$u_{21}(x, y) = \begin{cases} 2x^2 y, & \text{if } y \in [0, \tfrac{1}{2}], \\ -2(y-1)x^2, & \text{if } y \in (\tfrac{1}{2}, 1] \end{cases}$$
$$u_{22}(x, y) = (x + 1) \sin(\pi y).$$

The multicriteria game G has a single ideal Nash equilibrium in the point $(0, 0.5)$ with the corresponding payoffs of $(0.25, 0)$ [10].

The G game is composed by two unicriterial games: G_1 and G_2, where

$$G_1 = (\{1, 2\}, (S_i)_{i \in \{1,2\}}, (u_{1_1}, u_{2_1}))$$

and

$$G_2 = (\{1, 2\}, (S_i)_{i \in \{1,2\}}, (u_{1_2}, u_{2_2})).$$

Figure 1 depicts the detected strategies and payoffs for the Nash and Pareto equilibria for the single criterion game G_1. The Nash equilibrium consists of the solution $(0, 0.5)$ with payoffs of $(0.5, 1)$. The Nash equilibrium does not assure the highest payoffs, the Pareto equilibrium lies above the Nash equilibrium and spreads between $(0, 2)$ and $(1, 0)$.

Figure 2 depicts the detected strategies and payoffs for Nash and Pareto equilibria for the single-criteria game G_2. Similarly to the G_1 game, the Nash equilibrium consists of the solution $(0, 0.5)$ with payoffs of $(0.25, 0)$. Also, the Pareto equilibrium lies above the Nash equilibrium and spreads between $(0, 1)$ and $(1, 0)$.

The ideal Nash and multicriteria Pareto equilibria for the multicriteria game G is depicted in Fig. 3. The detected multicriteria Pareto-Nash and the ideal Nash equilibria are identical for the G game: the strategy profile $(0, 0.5)$ with a payoff of $(0.5, 1)$

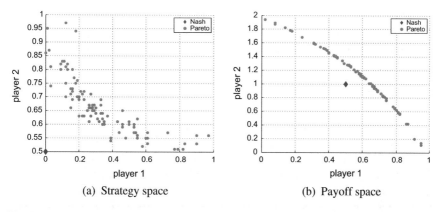

(a) Strategy space   (b) Payoff space

Fig. 1 The strategies and payoffs for the evolutionary detected Nash and Pareto equilibria for the G_1 game

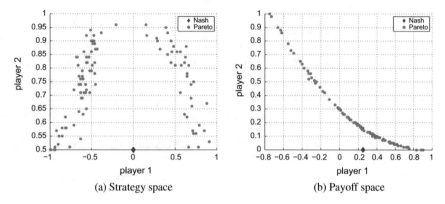

Fig. 2 The strategies and payoffs for the evolutionary detected Nash and Pareto equilibria for the G_2 game

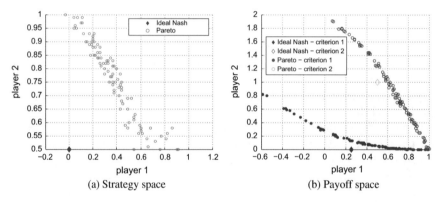

Fig. 3 The strategies and payoffs for the evolutionary detected Ideal Nash and multicriteria Pareto equilibria for the multicriteria game G

for the first criteria and $(0.25, 0)$ for the second criteria respectively. Note that the Pareto-Nash and the Ideal Nash equilibrium for G is the identical with the Nash equilibrium of the G_1 and G_2 games. The multicriteria Pareto equilibrium for game G is different than the Pareto equilibria for G_1 and G_2.

7 Conclusions

Standard Game Theory models players as rational agents whose only goal is to maximize their own payoffs. This is an unrealistic assumption since real-life players in most cases base their decisions on more then one, usually conflicting, criteria. Multicriteria games are extensions of standard non-cooperative games that allow vector payoffs. Thus multicriteria games model real-life situations more accurately. Surprisingly, multicriteria games have been neglected; several equilibria types have been defined but their detection did not receive much attention.

Inspired from multicriteria optimization we propose an evolutionary method for detecting equilibria in multicriteria games. This method is based on generative relations, that are efficient tools for comparing two strategy profiles. Most standard game equilibria can be characterized by generative relations [5, 6]. We define new generative relations for several multicriteria equilibria: Nash-Pareto equilibrium, Ideal Nash equilibrium and Pareto equilibrium. With the use of generative relations and evolutionary multiobjective optimization algorithms various multicriteria equilibria can be detected. For this purpose any state of the art multiobjective algorithm can be used.

In our numerical experiments we illustrate our method in case of discrete and continuous games. The results indicate the potential of the proposed approach.

We consider that the study of multicriteria games offers many challenging future possibilities. Based on multicriteria games new more realistic game theoretic models can be built in the area of economics, psychology, social sciences, etc. Future research direction is the extension of the proposed method for the detection of mixed strategies in multicriteria games. Also, future work involves the study of more complex multicriteria games with many players ($n > 3$).

References

1. Bade, S., Haeringer, G., Renou, L.: More strategies, more nash equilibria. J. Econ. Theory **135**(1), 551–557 (2009)
2. Borm, P.E.M., Tijs, S.H., van den Aarssen, J.C.M.: Pareto equilibria in multiobjective games. In: Technical report, Tilburg University (1988)
3. Borm, P., van Megen, F., Tijs, S.: A perfectness concept for multicriteria games. Math. Methods Oper. Res. **49**, 401–412 (1999)
4. Christos, H.: Papadimitriou: on the complexity of the parity argument and other inefficient proofs of existence. J. Comput. Syst. Sci. **48**(3), 498–532 (1994)
5. Dumitrescu, D., Lung, R.I., Mihoc, T.D.: Evolutionary equilibria detection in non-cooperative games. Applications of Evolutionary Computing, pp. 253–262. Springer, Heidelberg (2009)
6. Lung, R.I., Dumitrescu, D.: Computing nash equilibria by means of evolutionary computation. Int. J. Comput. Commun. Control **6**, 364–368 (2008)
7. McKelvey, R.D., McLennan, A.: Computation of equilibria in finite games. Handbook of Computational Economics, pp. 87–142. Elsevier, Amsterdam (1996)
8. Nash, J.: Non-cooperative games. Ann. Math. **54**(2), 286–295 (1951)
9. Osborne, M.J.: An introduction to game theory. In: Oxford University Press (2004)
10. Radjef, M.S., Fahem, K.: A note on ideal nash equilibrium in multicriteria games. Appl. Math. Lett. **21**, 1105–1111 (2008)
11. Shapley, L.S., Rigby, F.D.: Equilibrium points in games with vector payoffs. Nav. Res. Logist. Q. **6**(1), 57–61 (1959)
12. Storn, R., Price, K.: Differential evolution a simple and efficient heuristic for global optimization over continuous spaces. J. Glob. Optim. **11**(4), 341–359 (1997)
13. Thomsen, R.: Multimodal optimization using crowding-based differential evolution. In: IEEE Proceedings of the Congress on Evolutionary Computation, CEC'04, pp. 1382–1389 (2004)
14. Voorneveld, M., Grahn, S., Dufwenberg, M.: Ideal equilibria in non-coperative multicriteria games. In: Uppsala - Working Paper Series, 1999:19 (1999)
15. Wang, S.Y.: Existence of a pareto equilibrium. J. Optim. Theory Appl. **79**, 373–384 (1993)
16. Zhao, J.: The equilibria of a multiple objective game. Int. J. Game Theory **20**, 171–182 (1991)

A New Estimation of Distribution Algorithm for Nash Equilibria Detection

Tudor Dan Mihoc

Abstract One of the most popular solutions proposed for a noncooperative game is the concept of Nash equilibrium. An estimation of distribution algorithm is adapted to this problems' particularities in order to detect a sample of Nash equilibria for games in normal form. A generative relation is used to select the new strategy profiles that will generate the offspring. Several numerical experiments are conducted to validate the method for games with different numbers and types of equilibria.

1 Introduction

Computational **game theory** (GT) – extensively used in economics, social sciences, biology, engineering, computer science, and as well in philosophy – is the attempt to capture the agents' behavior in strategic situations, in which an individual's success in making choices depends on the choices of others [1].

Mathematical games are useful tools for modelling conflicting situations. A game is an unit formed by a set of players, a set of actions available to each player, and a set of payoff functions that each player aims to maximize.

Initially developed for zero sum games (competitions in which one individual does better at another's expense), GT has been extended to treat a much more wider class of interactions.

Game equilibria are the most common solutions proposed in GT. In order to provide adequate solutions, many equilibrium concepts have been developed. Among these proposed solutions, probably the most famous one is the Nash equilibrium [2].

Detecting game equilibria is a fundamental computational problem within noncooperative game theory, having connections with multi-criteria optimization.

Finding optimum methods of computing Nash equilibria is still one of the main aims in Computational game theory (CGT). There are many attempts to solve this

---

T. Dan Mihoc (✉)
Babeş Bolyai University, 1 Mihail Kogălniceanu Street, Cluj-Napoca, Romania
e-mail: mihoct@cs.ubbcluj.ro

© Springer International Publishing AG 2017   99
M. Emmerich et al. (eds.), *EVOLVE – A Bridge Between Probability,*
Set Oriented Numerics and Evolutionary Computation VII,
Studies in Computational Intelligence 662, DOI 10.1007/978-3-319-49325-1_5

problem, each having its own advantages but a general method that solve fast and reliably any game has not been developed yet.

The close relation between the estimation of distribution algorithms (EDA) and distributions of probabilities makes them natural tools for detecting Nash equilibria in mixed strategies profiles.

The paper is organized in five sections: First is the Introduction, the second presents some notions from game theory, and the third introduces the new method derived from the EDAs. Some numerical results are presented in the fourth section, and the fifth part contains the conclusions.

2 Prerequisites of Game Theory

In order to present the new method for equilibria detection, basic notions from game theory are presented in this section such as noncooperative games in normal form, pure strategies, mixed strategies, payoffs, utility functions, and strategy profiles. Also a generative relation will be applied in order to guide the EDA's algorithm type.

With respect to the relationship between the players' points of view, game theory (GT) can be divided in two major parts: cooperative game theory and noncooperative game theory. We will consider here the **noncooperative** game theory with **mixed strategies** solutions.

The players will also be rational, and they will have complete information on the game. This means that each player makes the best rational decision in order to achieve his/her goal (maximize the profit for example) and that every player has complete knowledge of the other players strategies, options, and payoffs.

A player's strategy space is the set of all strategies available to him. The set of strategies available to a player can be discrete (e.g., in Prisoners dilemma game) or continuous (like in the oligopolies of Cournot type).

A strategy profile (or simple "a strategy") is a complete plan of action for every stage of the game, regardless whether that stage actually arises in the play or not.

The payoff function for a player is a mapping from the cross product of players' strategy spaces to the player's set of payoffs, i.e., the payoff function of a player takes as its input a strategy profile and yields a representation of payoff as its output.

The games will be represented in normal form as a matrix for discrete strategies sets.

2.1 Game Definition

Following [1, 3], a game consists of a set of players (agents), and each player has a set of strategies available to her as well as a payoff function.

Definition 1 We consider a **finite strategic game** defined by $\Gamma = ((N, S_i, u_i), i = 1, \ldots, n)$ where

1. $N = \{1, \ldots, n\}$ the set of players, n is the number of players;

2. for each player $i \in N$, S_i represents the set of strategies available to him,

$$S_i = \{s_{i1}, s_{i2}, \ldots, s_{im_i}\}$$

where m_i represents the number of strategies available to player i and

$$S = S_1 \times S_2 \times \cdots \times S_n$$

is the set of all possible situations of the game and

$$(s_1, s_2, \ldots, s_n) \in S$$

is a pure strategy profile;
3. for each player $i \in N$, $u_i : S \rightarrow \mathbb{R}$ represents the payoff function.

We consider P_i be the set of real-valued functions on S_i. For the elements $p_i \in P_i$, we use the notation $p_{ij} = p_i(s_{ij})$. Let

$$P = \prod_{i \in N} P_i$$

and let

$$m = \sum_{i \in N} m_i.$$

We denote the points in P by $p = (p_1, \ldots, p_n)$, where $p_i = (p_{i1}, \ldots, p_{im_i}) \in P_i$.

In a mixed strategy profile p, an agent plays her available pure strategies with certain probabilities. The payoffs for the players that follow a mixed strategy profile is similar to the expected utility concept from decision theory:

$$u_i(p) = \sum_{s \in S_i} u_i(s) p(s)$$

where

$$p(s) = \prod_{j \in N} p_j(s_j)$$

2.2 Solution Concepts

The most basic premise of (noncooperative) game theory is that players are egoistic and rational [1]. This means that each player is primarily concerned about achieving his/her best possible payoff and assumes the same behavior on the part of all other players.

Each player's behavior is described by a strategy. Therefore, a player's strategy specifies what a player will do for each possible development of the game, not just the likely or rational ones. What a player will do, however, may involve probabilistic statements. There are three possibilities:

1. A pure strategy is the choice of one possible move at each of a given player's information sets with certainty. They are the sole choices that appear in the normal form;
2. A mixed strategy is a probability distribution over a player's pure strategies. Mixed strategies are usually involved in normal form game solutions;
3. A behavioral strategy is the specification of a probability distribution over the moves available at each information set of a given player.

Once each player has adopted a strategy, the resulting strategy profile determines the state of the game.

The basic solution concept of game theory is that of strategic equilibrium.

John Von Neumann, in 1928, [4] proposed the first modern definition and existence proof of strategic equilibrium in the case of zero-sum games. John Nash, in 1950, generalized the ideas to the nonzero-sum case and obtained an existence proof for what is now known as a Nash equilibrium [2].

A (mixed) strategy profile for the game Γ represents a Nash equilibrium if no player will increase her payoff by changing her own strategy while the others do not modify theirs.

John F. Nash proved that every game has a mixed strategies equilibrium. However, the proof is an existence proof (and uses a fixed point theorem), it does not give an explicit method to construct the equilibrium solution.

We also denote by (s_{i_j}, s_{-i}^*) the strategy profile obtained from s^* by replacing the strategy of player i with s_{i_j} i.e.,

$$(s_{i_j}, s_{-i}^*) = (s_1^*, s_2^*, \ldots, s_{i-1}^*, s_{i_j}, s_{i+1}^*, \ldots, s_1^*).$$

2.3 Nash Ascendancy Relation

In order to detect an equilibrium for a certain game Γ using evolutionary techniques, the search can be guided similarly to the detection of the Pareto set for a multi-objective optimisation problem. There are many similarities between the multi-objective optimisation problems and solving games.

A particularity of the games is, if we look from a multi-objective problem point of view, that the number of players equals the number of variables and the number of objectives.

A fitness solution for Nash equilibria detection using evolutionary techniques has been developed in [3].

Let us consider two pure strategy profiles s and s' from S. Let $k : S \times S \to \mathbb{N}$ be an operator that associates the cardinality of the set [5]

$$k(s, s') = |(\{i \in \{1, \ldots, n\}| u_i(s'_i, s_{-i}) \geq u_i(s), s'_i \neq s_i\}|$$

to the pair (s, s').

This set is composed by the players i that would benefit if given the strategy profile s would change their strategy from s_i to s'_i, i.e.,

$$u_i(s'_i, s_{-i}) \geq u_i(s).$$

Let $s, s' \in S$. The strategy profile s *Nash ascends* the strategy profile s' and we write $s \prec s'$ if the inequality

$$k(s, s') < k(s', s)$$

holds.

Thus, a strategy profile s ascends strategy profile s' if there are less players that can increase their payoffs by switching their strategy from s_i to s'_i than vice versa.

Remark 1 Two strategy profiles $s, s' \in S$ may have the following relation:

1. either s dominates s', $s \prec s'$ $(k(s, s') < k(s', s))$
2. or s' dominates s, $s' \prec s$ $(k(s, s') > k(s', s))$
3. or $k(s, s') = k(s', s)$ and s and s' are considered indifferent aka incomparable (neither s dominates s' nor s' dominates s).

This relation can be used as a relative measure of how "close" of Nash is a pure strategy profile in comparison with another one [6].

Proposition 1 $\prec$ *is the generative relation of the Nash equilibrium, i.e., nondominated strategies with respect to $\prec$ are the Nash equilibria of the game.*

Remark 2 The Nash-ascendancy generative relation in general is not transitive.

3 Estimation of Distribution Algorithm

An estimation of distribution algorithm (EDA) is an optimization technique that searches for potential solutions by sampling explicit probabilistic models of promising candidate solutions. It uses a population of potential solutions to the problem, starting with a random-generated population from all admissible solutions.

Using a fitness function, the population is evaluated and a numerical ranking is assigned to each individual. A subset of the most promising solutions are selected by a selection operator from this ranked population.

From this subset of selected, well-performing solutions, we estimate the probability distribution that characterizes it. Then, new solutions are sampled from this probability distribution and replace the previous population.

We repeat the process until some termination criteria are met (e.g., when the number of iterations reaches some threshold).

An EDA, in general, follows the next steps [7]:

1: **procedure** EDA
2:     Generate initial sample population of size M
3:     **while** termination criteria is not fulfilled **do**
4:         Select Q promising solutions where $Q \leq M$
5:         Calculate joint probability distribution of selected individuals
6:         Generate offspring according to the calculated probability distribution and replace parent.
7:     **end while**
8: **end procedure**

The similarities between EDAs and the mixed strategy concept from GT suggest a similar method for detecting Nash equilibria for noncooperative games. The main goal is to detect a n-tuple of distributions of probabilities (a Nash equilibria sample) using the fact that a mixed strategy profile is a solution that represents basically a set of pure strategy profiles.

We will adapt this algorithm to the specific requirements of our problem.

Instead of one discrete distribution of probability, we will have a vector of n distributions since a mixed strategy profile is composed of n such distributions.

A member of the population at this point will be naturally composed of n pure strategies, i.e., a pure strategy profile. Each player's strategy will be sampled according to his mixed strategy (his own distribution of probability).

In order to select the samples that will be used to compute the new mixed strategy profile, we use the generative relation for Nash equilibria. If a strategy profile dominates another, then this profile will be selected in the set that will be used to compute the new joint new mixed strategy profile. If the two strategy profiles are neutral with respect to each other, then both will be selected.

For each player, a joint probability distribution will be calculated using the selected strategies profiles. The result will be his mix strategy profile in the new iteration.

The modified algorithm will be

1: **procedure** MODIFIED EDA
2:     Random generate the initial mixed strategy profile p
3:     **while** termination criteria is not fulfilled **do**
4:         Generate the sample population of pure strategy profiles, of size M according with the probability distributions from p
5:         Select Q promising pure strategy profiles using the generative relation, where $Q \leq M$
6:         Calculate for each player the joint probability distribution of selected individuals
7:         Generate from these distributions the new mixed strategy profile p'.
8:     **end while**
9: **end procedure**

The algorithm will end after a number of iterations established a priori.

4 Numerical Experiments

For the numerical experiments, we consider several noncooperative games in normal form that have different numbers and types of equilibria (in pure and mixed strategies).

In all depicted cases, a Nash equilibrium is detected. For each example, 30 runs with different seeds for the pseudo-random generator were conducted. The number of pure strategies samples in the population is 100 for all experiments, and the number of iterations will be also 100.

4.1 The Matching Pennies Game

There are two players, each having a coin in the hand. Each player has two options: to put the coin down with the head or tail up. If both players put the coin with the same face down, the first player takes the two coins if not the second one will gain them.

For this game, there is no pure strategy Nash equilibrium. The only solution (equilibrium) in mix strategies $((0.5; 0.5)(0.5; 0.5))$ is detected with accuracy, in 30 runs with different random seeds, with a standard deviation of 0.0.

4.2 Rock-Paper-Scissors Game

Another famous game with one single solution in mixed strategies is Rock–paper–scissors (Table 1). In 30 runs of the algorithm, the average of the detected solutions is $((1/3; 1/3; 1/3)(1/3; 1/3; 1/3))$ with a standard deviation of 0.03.

4.3 Game with only Pure Strategies

Let us consider the game presented in Table 2. This game has two pure Nash equilibria, and the algorithm detects them.

Table 1 The payoffs for the Rock-paper-scissors game

	R	P	S
R	0, 0	−1, 1	1, −1
P	1, −1	0, 0	−1, 1
S	−1, 1	1, −1	0, 0

Table 2 The payoffs for a game with two Nash equilibria in pure strategies		1	2
	1	1, 0	3, 0
	2	0, 2	0, 1

At each run, the algorithm is biased towards the most close solution to the initial random-generated mixed strategy profile.

5   Conclusions and Further Work

Different equilibria, considered solutions in GT, can be characterised by generative relations between game strategy profiles. Binary generative relations for Nash equilibria are considered.

A new method for detecting a sample of Nash equilibria on mixed strategies based on the estimation of distribution algorithms has been proposed.

The EDA algorithm is modified to detect a good approximation of a vector of discrete distributions of probabilities using as candidate solutions profiles of pure strategies.

The selection operator is based on Nash equilibria domination relation that can be used as a quality measure for a strategy profile.

The numerical experiments that are conducted underline the potential of this method. Further work will imply a more robust algorithm, capable of detecting more than one equilibria sample.

Acknowledgements This work was supported by a grant of the Romanian National Authority for Scientific Research and Innovation, CNCS UEFISCDI, project number PN-II-RU-TE-2014-4-2560.

References

1. Osborne, M.J., Rubinstein, A.: A Course in Game Theory. MIT Press, Cambridge (1994)
2. Nash, J.F.: Non-cooperative games. Ann. Math. **54**, 286–295 (1951)
3. Lung, R.I., Dumitrescu, D.: Computing nash equilibria by means of evolutionary computation. Int. J. Comput. Commun. Control **III**, no. suppl.issue, 364–368 (2008)
4. von Neumann, J., Morgenstern, O.: Theory of Games and Economic Behavior. Princeton University Press, Princeton (1994)
5. Gaskò, N., Dumitrescu, D., Lung, R.I.: Evolutionary detection of Berge and Nash equilibria. Nature Inspired Cooperative Strategies for Optimization (NICSO 2011), pp. 149–158. Springer, Berlin (2012)
6. Nagy, R., Suciu, M., Dumitrescu, D.: Lorenz equilibrium: equitability in non-cooperative games. In: Proceedings of the Fourteenth International Conference on Genetic and Evolutionary Computation Conference. ACM (2012)

7. Kern, S., Müller, S.D., Hansen, N., Büche, D., Ocenasek, J., Koumoutsakos, P.: Learning probability distributions in continuous evolutionary algorithms - a comparative review. Nat. Comput. **3**(1), 77–112 (2004)
8. McKelvey, R.D., McLennan, A.: Computation of equilibria in finite games. Handbook of Computational Economics, pp. 87–142 (1996)

Part III
Algorithm Design for Real World Challenges

Multi-objective Optimisation
by Self-adaptive Evolutionary Algorithm

John M. Oliver, Timoleon Kipouros and A. Mark Savill

Abstract Evolutionary algorithms (EAs) have been used to tackle non-linear multi-objective optimisation (MOO) problems successfully, but their success is governed by key parameters which have been shown to be sensitive to the nature of the particular problem, incorporating concerns such as the numbers of objectives and variables, and the size and topology of the search space, making it hard to determine the best settings in advance. This work describes a real-encoded multi-objective optimising EA (MOOEA) that uses self-adaptive mutation and crossover, and which is applied to optimisation of an airfoil, for minimisation of drag and maximisation of lift coefficients. The MOOEA is integrated with a Free-Form Deformation tool to manage the section geometry, and XFoil which evaluates each airfoil in terms of its aerodynamic efficiency. The performance is compared with those of the heuristic MOO algorithms, the Multi-Objective Tabu Search (MOTS) and NSGA-II, showing that this GA achieves better convergence.

1 Introduction

Genetic Algorithms (GA) are a class of evolutionary algorithm (EA) originally proposed by Holland [20] and expanded upon by Goldberg [18] and Schaffer [32], that are heuristic, stochastic methods of searching very large non-linear problem spaces. EAs are used in particular for global optimisation problems upon which classical optimisation methods do not perform well, in order to attempt to obtain optimal or near optimal solutions [23].

GAs are characterised by populations of potential solutions that converge towards local or global optima through evolution by algorithmic selection as inspired by neo-Darwinian [7] evolutionary processes. An initial population of random solutions is created and through the evaluation of their fitnesses for selection for reproduction, and by the introduction of variation through mutation and recombination (crossover),

J.M. Oliver (✉) · T. Kipouros · A.M. Savill
School of Engineering, Cranfield University, College Rd, Cranfield MK43 0AL, UK
e-mail: j.m.oliver@cranfield.ac.uk

© Springer International Publishing AG 2017
M. Emmerich et al. (eds.), *EVOLVE – A Bridge Between Probability,*
Set Oriented Numerics and Evolutionary Computation VII,
Studies in Computational Intelligence 662, DOI 10.1007/978-3-319-49325-1_6

the solutions are able to evolve towards the optima. It should be noted that GAs are not attempting to mimic evolution as it occurs in the biological domain, since evolution is directionless in the absolute sense, and optimisation is by definition seeking a goal, rather they are using simplified mechanisms which have been shown to work in optimising processes.

Research into EAs over more recent years by Fleming [16], Fonseca [17], Deb [8], Zitzler [46] and others, has extended their use to multi-objective optimisation problems (hence MOOEA), in which two or more conflicting objectives, each with their own criteria, are optimised simultaneously, with the goal of yielding a Pareto-optimal attainment surface from which a trade-off solution can be chosen by a higher-level decision maker.

GA performance on a given problem has been shown, since De Jong [24], to be extremely sensitive to the settings of its parameters, these being the probabilities of mutation and crossover occurring, the population size and the number of players in a tournament selection (when this selection method is used). Moreover, for certain real/continuous encoded GAs, it is necessary to consider operators' polynomial distribution indices [9].

Real-encoded GAs can be thought of as being similar to Evolutionary Strategies (ES) introduced by Rechenberg and Schwefel as described by Bäck [4], except that ESs also are able to self-adapt their control parameters (or strategy parameters as they call them). The GA described in this work adopts this extra capability. The term *self-adaptive* used here is meant in the sense of that coined by Eiben et al. [14], to indicate control parameters of the GA that are encoded in the chromosome along with the problem definition parameters applying to the objective functions (the *main* parameters), and that these control parameters are subject to change along with the *main* parameters due to mutation and crossover. This is different from a purely *adaptive* control parameter strategy as in that case the change is instigated algorithmically by some feedback at the higher level of the GA rather than the lower level of each chromosome/solution in the population. The *deterministic* approach is rule-based and is not considered adaptive.

Eiben et al. [13] showed how population size and tournament selection size can be made to be self-adaptive, although in the former case to the detriment of performance of the optimisation. Nonetheless, the latter case was shown to improve performance, and the method by which a parameter whose context is the population can be set through the aggregation of its representation at the individuals within the population, can be extended to other parameters having the same high-level context. However, the above work only uses mutation to affect each self-adapting parameter gene, rather than including the parameter genes in the crossover of the chromosome as a whole, and the model used is a steady-state GA (SSGA) with relatively low replacement strategy rather than a generational one (GGA). This work uses a fixed tournament size in order to keep all the self-adaptation occurring at the level of the individual, rather than by aggregation, since this is the focus of the work.

Zhang and Sanderson [42, 43] describe differential evolution (DE) algorithms that use self-adaptation, including their multi-objective (MO) JADE2 and JADE algorithms, that generate new values for mutation factors and crossover probabilities based on probability distributions governed by self-adapting *means*. DEs [38] are similar to GAs but new solutions are produced by adding the weighted difference of two population vectors to a third, to create a new donor vector which is recombined (crossover) with a target (parent) vector to produce the trial (child) vector. Differences between GAs and DEs, both algorithmic and from a performance perspective, are discussed in [41]. In a DE scheme, the mutation factor is a weight rather than a probability as in a GA, and notably crossover acts on whole parameters (the genes in a GA) rather than parts of parameters (Holland's *schemas*).

Sareni et al. [31] describe self-adaptation in a multi-objective genetic algorithm (MOGA) in which there is a self-adaptive choice between three different crossover operators for crossover, and in which mutation is self-adapted by the standard deviation of the amount of perturbation applied to a gene. Both of these mechanisms are different to the ones employed by the MOOEA in this work.

Tan et al. [40] expounded their binary MOGA in which the mutation rate is deterministically assigned as a function of time, and Tan et al. [39] discuss a deterministic binary MOGA in which rules assign values for mutation and crossover probabilities. Ho et al. [19] used a binary GA for single objective optimisation in which sub-population groups adapted their mutation or crossover rates based on feedback from average fitness increase, while Li et al. [29] investigated diversity-guided mutation and deterministically adaptive mutation and crossover rates in a binary single-objective GA. These works all found their implementations of the various adaptive methods provided advantages on mathematically based benchmark problems, but differ from this work which is concerned with self-adaptation in a real-encoded multi-objective GA that addresses a real-world optimisation problem having time-consuming function evaluations.

This work presents a real-encoded generational MOOEA employing elitism in which each solution has an evolving self-adaptive mutation-rate and self-adaptive crossover-rate, together with their own perturbation factors, encoded in its chromosome and which are subject to both mutation and crossover themselves, along with the *main* problem parameters. The MOOEA is used on a real engineering multi-objective optimisation problem (MOOP), that of airfoil optimisation, and its performance on the problem is compared with two other leading heuristic algorithms, Multi-Objective Tabu Search (MOTS) and NSGA-II. The MOOEA uses a novel crossover mechanism in order to recombine both the mutation rate and crossover rate control parameters at the level of the chromosome, and unlike other GAs, controls the number of duplicate chromosomes in each generation. Whereas DE algorithms have used probability distributions governed by self-adapting *means*, this MOOEA uses its own mutation and crossover operators to control its self-adapting control parameters in the same way that they change the *main* genes.

2   A Self-adaptive Multi-objective Evolutionary Algorithm

Ganesh, the MOOEA developed in this work, was inspired by the NSGA-II algorithm [10] with some modifications, to: initialisation of population and solution, the
non-domination sorting method, the construction of the new generation, the addition
of repairable Hard Constraints, the adoption of a plug-in architecture, and of course
the self-adaptive aspect. Soft constraints are implemented similarly to NSGA-II,
requiring the constraint definition to return an increasingly negative number indicating the increasing degree of violation, and where 0 indicates no violation. Internally
the MOOEA is constructed to minimise, requiring objective functions that maximise
to return a negative number, by the principle of duality [8]. A tournament selection
method of degree two is used, polynomial mutation [11] is used along with a simulated binary crossover (SBX) [9] for real parameters, and the crossover strategy used
in the problem discussed here is uniform crossover, although a multi-locus gene-swap
crossover is available also. Self-adaptive crossover requires further consideration and
is discussed more fully, below.

The non-domination sorting is amended from NSGA-II to ensure that each solution is compared with every other one once in a simple and efficient manner as
given below, with the number of comparisons being of the same order, $O(mN^2)$,
as that of the *continuously updated* method [8], and the actual number of comparisons made will on average be the same. Here, *compare* solutions means perform the
$(z^1 \succ z^2)$ dominance test [46], meaning that z^1 is not worse than z^2 in all objectives
and better in at least one objective. The method of updating dominated-by count and
dominated-solutions lists are modified accordingly.

Where there is a population p having n solutions:

```
for   p in 0 to n-1
    for   q in (p + 1) to n-1
        compare q to p
```

or more formally:

$$\forall p \in \{0, \ldots, n-1\}, \forall q \in \{p+1, \ldots, n-1\} : (q \succ p) \to q$$

The new generation is produced by pruning one solution at a time from the merged
parent and child populations and recalculating the distance/crowding metric each
time, giving a more accurate estimate of the best solution to remove with respect to
the crowding (and non-dominated ranking) metric, a method which was later found
to have been already tried by [28].

This MOOEA additionally provides the ability to choose the cardinality of duplicate solutions in each generation, meaning that 0, 1 or many duplicates may be kept,
with the default being many. Zero duplicates means one solution having no duplicates,
and so on, where a duplicate is defined as all corresponding genes in both chromosomes having the same values. Duplicates can arise even in real-encoded problems
which are not combinatoric, due to elitism. The ability to control the existence of

duplicates is achieved here through the use of a linked hash map data structure where the key is the chromosome and the value includes a count and list of chromosomes having the same genes.

Hard constraints may be either of pre- or post-evaluation types, where pre-evaluation hard constraints are allowed to be repairable, whereas post-evaluation hard constraints must cause discarding of failing solutions. Repair is effected by changing the parameters (the *genes*) until the solution is once again within the constraint. Since repair occurs before evaluation of the solution (the determining of the objective function values), repaired solutions are available in the current generation as normal. Of course hard constraints which rely upon the values of objective functions as part of their violation detection, may only be of the post-evaluation type. Soft constraints may be of either type, with the same proviso, but do not provide a discard option.

Initialisation at both population level and solution level have defaults which are able to be over-ridden by the problem definition, enabling pre-defined data to be included, and alternative distribution functions to be used.

The plug-in architecture of the algorithm enables the optimisation problem to be specified separately as a new code module, thus each new optimisation problem adds code, rather than requiring changes to the existing code base, and enables the optimisation problem to be effectively a parameter of the algorithm.

2.1 Self-adaptation

Bäck et al. [5], use a random mutation rate initially, each solution being initialised to a random number in the range (0.001, 0.25), however he suspected that this randomness slowed down convergence to some extent. Ganesh allows mutation and crossover rates to be specified for the initial population, or to be set to random values in a uniform distribution, or to default to certain values. The default mutation rate of each solution would be set to 1/n where n is the number of variables of the objective functions (OFs), and the default crossover rate would be 0.6, both as probability of occurrence. This MOOEA (henceforth referred to as a GA for brevity) also allows alternative initialisers to be written and specified per problem, allowing for different probability distributions, such as the uniform or Gaussian, however this work uses the uniform distribution.

Similarly to Bäck [4] and Smith and Fogarty [37] (a steady-state GA), mutation first occurs to the gene encoding the mutation rate and then the new mutation rate is applied to the *main* genome, but unlike the previous studies, this is based on a generational GA, that is one in which the entire population is in theory able to be replaced by fitter solutions, and for which the variables, and operator parameters, are encoded as real numbers in the genes.

The GA control parameters undergoing self-adaptation are the mutation probability pM (per gene) and the crossover probability pC (per chromosome), and also the associated polynomial distribution indices, [9, 11], for each, ηM and ηC respectively,

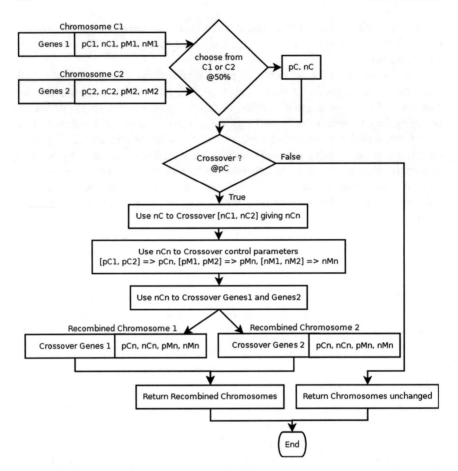

Fig. 1 Flow-chart explaining self-adaptive crossover

which are all real values. Each solution has a chromosome encoding its objective function parameters and its control parameters. Mutation occurs to all of the parameters including the control ones and their indices, but mutation occurs first to the control ones at the current rate of mutation, and then the *main* ones using the newly mutated values.

The uniform crossover specifies that each gene has a 50% chance of crossing over if the chromosome is to undergo crossover at all, and the probability of chromosome crossover occurring is given by pC. However since crossover occurs between chromosomes but each chromosome has its own pC, the pC to be used is chosen stochastically at 50% probability from either of the parent chromosomes selected for breeding, and the ηC is taken from the same chromosome. The ηC value is then used in the crossover of the respective controls from each chromosome (pC, ηC, pM and ηM) and the *main* chromosome genes, with the new control values being written to the recombined chromosomes, as shown in Fig. 1.

3   Airfoil Optimisation

The real-world engineering problem to which this GA is applied is airfoil optimisation, using the NACA 0012 airfoil section, as previously carried out by Oliver et al. [30], following on from Kipouros et al. [27] in which Multi-Objective Tabu Search (MOTS) software [22] and NSGA-II were used. NACA 0012 [1], Fig. 2, is a standard symmetric airfoil having a 12 % thickness to chord length ratio, defined originally by the U.S. National Advisory Committee for Aeronautics, now part of the National Aeronautics and Space Administration (NASA). Airfoil shape modification is carried out by free-form deformation (FFD) [33] code and the shape is evaluated for aerodynamic efficiency by Drela's XFoil tool [12] which calculates moment, drag and lift coefficients of flow, based on eight parameters as illustrated in Fig. 3.

   The MOOEA with Xfoil had previously [30] been found to locate extreme minima that give rise to unfeasible airfoil shapes, as under certain conditions Xfoil would not converge and would not feedback the convergence failure, hence the unfeasible coefficients were still selected for by the MOOEA, and would have an inappropriately high fitness, ensuring they remained in final results. Soft constraints had been applied

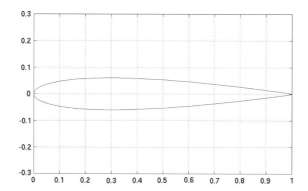

Fig. 2   The NACA 0012 airfoil [34]

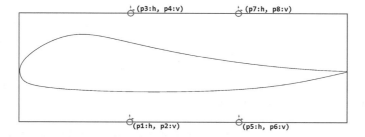

Fig. 3   An airfoil is a cross-section of a wing and is shown here enclosed in its free-form deformation hull with the eight deformation parameters, comprising *four control points* with a *horizontal* and *vertical shift*, that define its shape altering. Kipouros et al. [27]

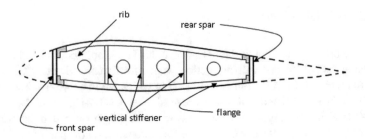

Fig. 4 An airfoil showing strengthening spars and vertical stiffeners

to both lift and drag coefficients in an attempt to minimise this problem, and this had improved the performance yet not eliminated the problem entirely.

In order to preclude the necessity of removing unfeasible designs from future result sets, the Xfoil software was further enhanced to ensure that convergence problems are identified to the MOOEA by assigning extremely low-quality fitness values for each objective function of any candidate solution for which convergence in Xfoil is a problem. The MOOEA is therefore able to eliminate these unfeasible designs through its normal selection process. It should be noted that Xfoil was originally designed to be an interactive tool, and the challenges found using it in batch mode were not due to the original author.

The airfoil is subject to two hard geometrical constraints, these being implemented inside XFoil, parametrically: the thickness of the airfoil section at (a) 25 % and (b) 50 % of the chord as defined by NACA-0012, which ensure there is a minimum volume in which to place strengthening spars towards the leading and trailing edges, thus discovered optimised designs should in theory be feasible and practicable, as in Fig. 4.

The optimisation definition ensures that each candidate design has the same angle of attack, Fig. 5, so that the objective function results are comparing equivalent measurements, by choosing FFD parameters which do not alter the position of the leading and trailing edges of the airfoil or the chord.

The FFD's eight design parameters are encoded in the GA as real numbers in the genes of each solution's chromosome, and the FFD code modifies the airfoil relative to a given datum design vector defining the geometry, based on the parameters from the GA design vector. FFD expresses the modified geometry as sets of x–y coordinates in a form that XFoil can receive, XFoil then calculates the coefficients of moment, drag (C_D), and lift (C_L) of the modified geometry and returns the latter two results, C_D and C_L, to the GA. Since the goal of this work is to optimise the airfoil with respect to drag and lift as a bi-objective problem, the coefficient of moment is not used at this time.

The objective functions OF1 and OF2, as given by Eqs. (1) and (2), define maximisation of the lift coefficient and minimisation of the drag coefficient respectively, normalised by their respective datum values. The datum values, ($C_L = 1.46444$, $C_D = 0.0305108$), are the original coefficient values of the standard airfoil section.

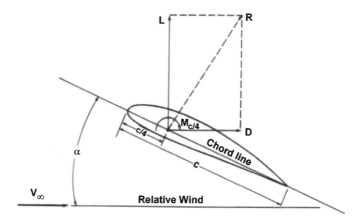

Fig. 5 Diagram showing α the Angle of Attack, lift (L) and drag (D) vectors and the airfoil chord [2]

$$OF1 = \max F(C_L) = -\frac{C_L}{C_{L,\text{datum}}} \tag{1}$$

$$OF2 = \min F(C_D) = -\frac{C_D}{C_{D,\text{datum}}} \tag{2}$$

Ganesh is set to perform 6,000 function evaluations in each run with a population size of 120 and being allowed to run for 50 generations, as was performed in [27]. For all cases here, Ganesh is set to allow no duplicates. The probability of crossover pC is initially set to 0.9 and that of mutation pM to 0.5 for each member of the initial population, and their respective polynomial indices ηC to 10 and ηM to 20, as was the case for NSGA-II, but in the succeeding generations these values self-adapt. The probabilities may self-adapt in the interval [0, 1] while the polynomial indices may self-adapt in the interval [1, 100], noting that for the latter, larger values cause smaller perturbations in the original gene values, and vice-versa.

The problem definition used by the GA also defines the range by which the design vector is allowed to be modified, and thus how much the geometry of the airfoil may change, specified as follows: ± 0.3, ± 0.4, ± 0.6 and ± 1.0, with a larger range enabling larger variation in the free-form deformation. A given run of the GA uses one of those ranges, and the corresponding results are compared with those of MOTS and NSGA-II. It should be noted that the *range* specifies by how much one of the 8 FFD geometrical parameters is permitted to change - it is not a fixed amount, and the lower ranges are by definition therefore encapsulated in the higher ranges. The value of each parameter is relative to the geometry and is dimensionless. Large deformations give rise to more exotic airfoil shapes which are more likely to be impractical in reality and which tend to be more demanding, or even impossible, to compute, thus the ranges used enable useful comparisons to be made in an escalating fashion over the objective space.

4 Results

Although 80 runs (20 per range) of the GA had been performed for the original work [30], automatic removal of unfeasible designs had not been possible, therefore a statistical analysis of the results would have provided little value and may have been potentially misleading.

These final results are the consequence of integrating the MOOEA with the new version of Xfoil, furthermore, the other algorithms, MOTS and NSGA-II, have also been re-run with the new Xfoil, enabling a thorough statistical analysis and comparison of them all to be performed, as is presented here. As before, another set of 20 runs per range per algorithm is performed, giving a total of 80 runs per algorithm and therefore 240 runs overall.

Figures 6, 7, 8, 9, 10 and 11 show scatter plots of non-dominated solutions in the objective space obtained by Ganesh and the other algorithms, in which OF1 and OF2 give the normalised values of C_L and C_D, plotted along the x and y axes respectively, as previously described. All solutions found are considered feasible designs and none have been removed, and dominated solutions are not shown. The values of C_L are shown as negative since it is being maximised and the GA is constructed internally to assume minimisation. All results are for generation 50 (numbered as 0 to 49) unless

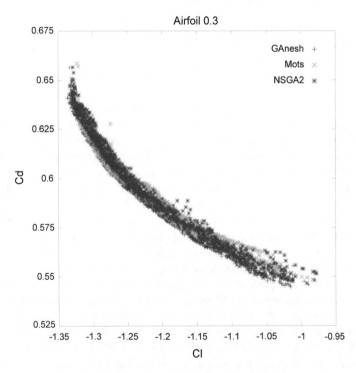

Fig. 6 Results for range ±0.3 showing Ganesh, MOTS and NSGA-II

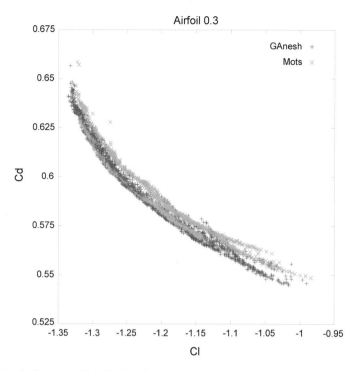

Fig. 7 Results for range ±0.3 showing Ganesh and MOTS only

stated otherwise in the figure caption, to provide a direct comparison with the MOTS and NSGA-II results obtained previously.

The number of generations for which the algorithms are permitted to run for is a limitation set in the scenario as a basis for comparison; Ganesh had not finished converging, as there were found to be dominated solutions in the last generations, thus if it had been permitted to run longer, even better results would have been obtained. This is likely to be true of the other algorithms too.

The PISA package [6] was used with the results obtained to produce standard metrics for hypervolume indicator [44] and ε-indicator [46], to understand the performance better through its statistical analysis and performance package. Table 1 gives the results of Kruskal–Wallis non-parametric one-tailed tests comparing each algorithm's 20 samples results (for a given range) against each of the others, in which the null hypothesis, $H0$, is that any variation seen between any two algorithm performances is due to random fluctuation within normal bounds, assuming an alpha value of 5 %.

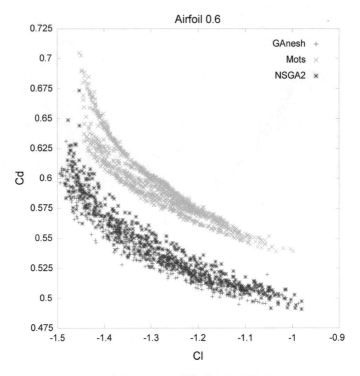

Fig. 8 Results for range ±0.6 showing Ganesh, MOTS and NSGA-II

When $H0$ is not rejected, one algorithm cannot be said to out-perform the other, conversely when $H0$ *is* rejected, the test suggests that the first algorithm does outperform the second, for the indicator under consideration. Table 1 gives the test results showing p-values representing the probability that the difference in performance seen is no greater than would be expected if the samples had the same means, thus where a p-value is less than 0.05, it is reasonable to reject $H0$ and say the first algorithm does out-perform the other. Where the p-value is greater than 0.05, $H0$ is not rejected and the algorithms probably have similar performance. The tests use the *hypervolume* indicator [44, 45], which was originally described as "the size of the space covered", and the ε-indicator [46], for which the means and standard deviations of the values used in the tests are given in Tables 2, 3 and 4.

The hypervolume indicator is a measure of how much area or (hyper-)volume, depending on the dimensionality of the objective space, is covered by an approximation front, thus it is not only an indicator of convergence but also of breadth of front. The ε-indicator (epsilon) is a measure of the minimum distance of translation needed to move every solution in the discovered front, so that the front weakly dominates the most converged front found. So for a given range, the most converged front from all 60 samples is chosen, by PISA, as the reference set against which the ε-indicator is measured.

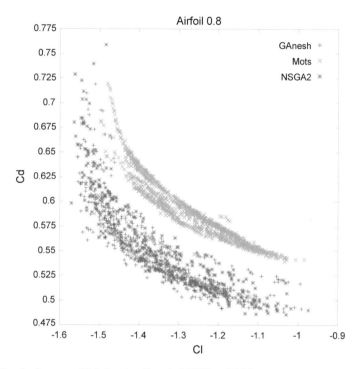

Fig. 9 Results for range ±0.8 showing Ganesh, MOTS and NSGA-II

The Kruskal–Wallis tests show that for these set of results, for both indicators across all ranges, Ganesh seems to out-perform MOTS in terms of convergence and has a similar breadth of front, although at the greater ranges, not as dense as MOTS. As can be seen in the scatter plot for the 1.0 range, MOTS does achieve several very good non-dominated points. The tests also show that Ganesh out-performs NSGAII for *hypervolume* at 0.3, both indicators at 0.6 and for ε-indicator at 0.8, but neither at 1.0. NSGAII on the other hand does not seem to out-perform Ganesh for any indicator at any range, while it also seems to outperform MOTS across indicators and ranges.

The means and standard deviations for the indicators given in Tables 2, 3 and 4 show that Ganesh tends to have a higher variation in indicator value than NSGA-II, and this seems to be borne out by the scatter plots, although Ganesh does seem to get solutions to the front edge of the Pareto plots. NSGA-II seems to achieve wider fronts at 1.0, while not seeming as well converged, which seems to be the reason that neither can be said to out-perform the other in all respects as shown by the Kruskal–Wallis tests.

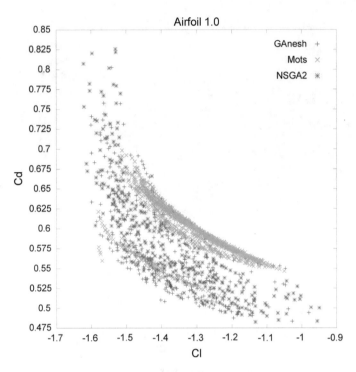

Fig. 10 Results for range ±1.0 showing Ganesh, MOTS and NSGA-II

Figure 6 has the results of NSGA-II plotted last, hence this set somewhat obscures the underlying results of MOTS and Ganesh, so the latter two are shown together in a separate plot, in Fig. 7.

Figure 11 is a repeat of Fig. 10 but with a much later generation of a Ganesh run: generation 863, which was the first generation in which all 120 solutions of the population of that generation were non-dominated. Prior to that point, every preceding generation had at least one dominated solution in it. As one would expect, this front dominates the others, but it also has quite a dense and wide front. It is reasonable to suppose that even this late generation is not yet the best performance that it might achieve, given that it has only just eliminated the last dominated solution(s). Figure 11 also shows, by means of an arrowed line, the airfoils for a number of selected points. The drag coefficient scale has been extended to show the datum airfoil point, towards the right at the top. The airfoils are chosen from each end of the approximate Pareto front and from near the middle, to show the response of aerodynamic performance to the shape as it is specialised away from the NACA012 airfoil.

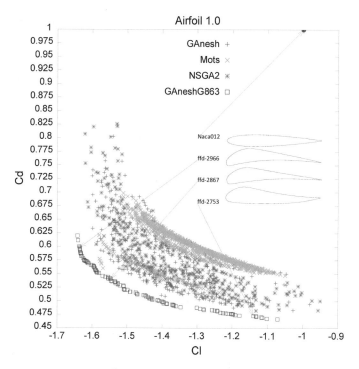

Fig. 11 Results for range ±1.0 showing Ganesh, MOTS and NSGA-II, and GaneshG863 which is the first generation of Ganesh having only non-dominated solutions (in generation 863), and a selection of resulting airfoils

An airfoil from generation 863 is shown in Fig. 12, which is near the middle of the Pareto front approximation. This represents a compromise design having good lift and good drag coefficient values. The graph above the airfoil section shows pressure coefficient distributions for the airfoil surfaces and boundary conditions. The parallel coordinates plot of Fig. 18 shows the design parameters for the selected approximate cL of this airfoil.

In Figs. 13 and 14 the means of the GA control parameters are plotted since each of the 120 solutions in each generation has its own value for each of these parameters. It can be seen that as the GA progresses through its generations, both pM and pC become smaller, hence the disturbance to good solutions is lessened, while their respective polynomial distribution indices become larger, which decreases the perturbation to the section geometry, thus at the start the GA is better at exploring the search space while towards the end it is better at converging to good solutions.

Figures 15 and 16 show how the standard deviations of the control parameters vary across the generations. As can be seen, they start off low, since the parameters are set at the start of the run, then increase rapidly at the early stages, but begin to increase less rapidly at around the middle of the run, with expectations of levelling out towards the end of the run. This is indicative of much variation across individual

Table 1 Kruskal–Wallis tests results for 3 independent data sets, for the the ε- and hypervolume indicators (ε and h respectively), comparing 20 sample runs per range for each of Ganesh (G), MOTS (M) and NSGA-II (N), showing p-values for $\alpha = 0.05$

Rng	Ind	G > M	G > N	N > M	N > G	M > G	M > N	Best	2nd-Best
0.3	ε	2.76E-05	0.61846	9.64E-06	0.38154	0.99997	0.99999	G/N	M
0.3	h	8.99E-15	0.00245	5.56E-10	0.99755	1	1	G	N
0.6	ε	3.20E-17	0.00074	6.17E-12	0.99926	1	1	G	N
0.6	h	2.90E-18	0.00009	5.61E-12	0.99991	1	1	G	N
0.8	ε	1.44E-12	0.03461	1.70E-09	0.96539	1	1	G	N
0.8	h	1.24E-12	0.07296	3.43E-10	0.92704	1	1	G/N	M
1.0	ε	7.74E-08	0.26335	8.31E-07	0.73665	1	1	G/N	M
1.0	h	4.33E-07	0.47693	5.37E-07	0.52307	1	1	G/N	M

Table 2 Means and standard deviations (SD) of the ε- and *hypervolume* indicators provided by PISA for Ganesh results

Ganesh

Range	Epsilon		Hypervolume	
	Mean	SD	Mean	SD
0.3	0.075457	0.039082	0.050667	0.023536
0.6	0.080692	0.033401	0.075564	0.032801
0.8	0.111012	0.032667	0.131357	0.044737
1.0	0.146081	0.056356	0.190044	0.101289

Table 3 Means and standard deviations (SD) of the ε- and *hypervolume* indicators provided by PISA for MOTS results

MOTS

Range	Epsilon		Hypervolume	
	Mean	SD	Mean	SD
0.3	0.155269	0.066874	0.122358	0.03185
0.6	0.317691	0.055242	0.401264	0.037391
0.8	0.251978	0.058163	0.387173	0.085911
1.0	0.234526	0.044423	0.351916	0.094109

Table 4 Means and standard deviations (SD) of the ε- and *hypervolume* indicators provided by PISA for NSGA-II results

NSGA-II

Range	Epsilon		Hypervolume	
	Mean	SD	Mean	SD
0.3	0.069424	0.026199	0.069150	0.015704
0.6	0.102902	0.021668	0.115390	0.029101
0.8	0.128359	0.028219	0.144089	0.042932
1.0	0.154388	0.037153	0.180101	0.046149

candidate solutions, for these controls, that nevertheless seems to reach a maximum level, perhaps a "requisite variety". The trends of the means do however seem to indicate that overall the solutions are evolving to better states.

Figure 17 is a plot that shows results in the form of parallel coordinates (‖-coords), the technique devised by Inselberg [21] and later used in the field of optimisation by Fleming [16], Siirtola [35, 36], and in engineering design by Kipouros et al. [25, 26], in which each dimension is oriented parallel to the others, thus transforming an n-dimensional point into a 2-dimensional polygonal line that relates the values in each dimension. This approach enables highly multi-dimensional data to be plotted uniquely and without loss of information, and here the entire design space of each

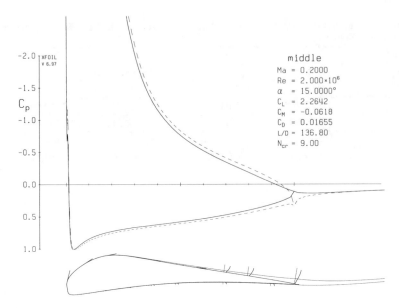

Fig. 12 An airfoil (ffd-2867) from range ±1.0 in generation 863, cL 1.547 cD 0.540 (the values in the figure itself are not normalised.), showing analysis of aerodynamic performance

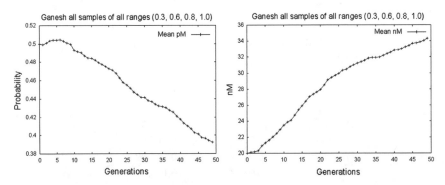

Fig. 13 Trends of the means of the pM and ηM control parameters against generation number for all ranges for all samples for Ganesh

solution, 8 variables and 2 objective function results, are plotted together. The plots were produced using the Parallax tool [3].

Thus Fig. 17 shows the eight parameters and two objective functions (Cl and Cd) of the design vector of all solutions in generation 863 of a run for range ±1.0, as shown in Fig. 11. This shows parameter 6 has been selected for value 1, the value that seems to achieve the greatest lift, while the lowest drag has also been selected, showing that the least drag corresponds to the least lift (as lift here is a negative amount as explained previously). As might be expected, the opposite value of parameter 6 is selected for least drag, while high values of p5 and p3 are also selected, both of

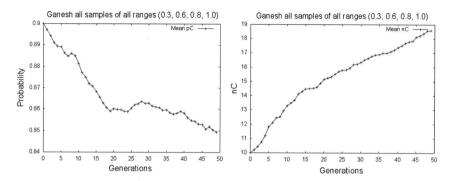

Fig. 14 Trends of the means of the pC and ηC control parameters against generation number for all ranges for all samples for Ganesh

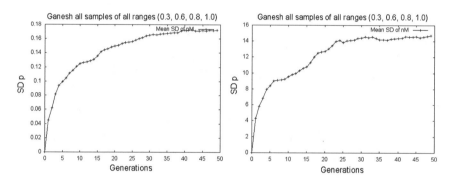

Fig. 15 Trends of the standard deviations of pM and ηM control parameters against generation number for all ranges for all samples for Ganesh

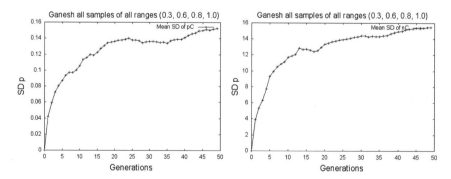

Fig. 16 Trends of the standard deviations of pC and ηC control parameters against generation number for all ranges for all samples for Ganesh

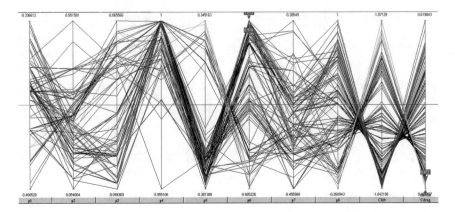

Fig. 17 Parallel coordinates plot showing Ganesh results for range ±1.0 at generation 863, which is the first generation having only non-dominated solutions

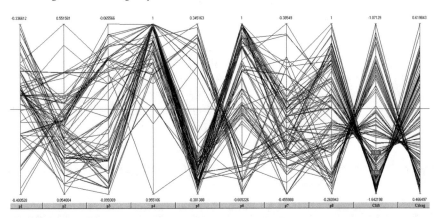

Fig. 18 Parallel coordinates plot showing Ganesh results for range ±1.0 at generation 863, with the airfoil cL and cD, see Fig. 12, selected

which are antagonistic for lift. Interestingly, p4 seems to have high values for both lift and drag. Figure 18 shows the plot for the selected airfoil of Fig. 12, as previously described.

5 Conclusion

Recalling that the range of deformation allowed in the FFD transformation is a dimensionless relative measure against a particular geometry, it was apparent that as the range increased, the MOOEA was able to find attainment surfaces that were better approximations of the Pareto-optimal front, as do the other algorithms, as it is intrinsically enabled to explore wider areas of the search space at earlier times. XFoil can take longer to run with larger variations in range as it may find it harder to converge successfully and indeed may fail to converge.

The ability of the GA to self-adapt its crossover and mutation rates seemed to help it improve exploration and convergence, since not being fixed, the rates are more likely to be appropriate at a given generation as they co-evolve along with the fitness of solutions. Although each new self-adaptive parameter can also be thought of as a factor increasing the *total* decision search space, the increase is by a relatively small percent. There is inevitably a trade-off between the impact on search efficiency by the small expansion of the search space through the addition of the control parameters, and the exploitation effect of possibly more appropriate control parameters arising. Moreover, the number of generations allowed in these experiments can be thought of as quite low, and at greater generations, the impact of self-adaptivity would be expected to be greater.

Specifying that zero duplicates are permitted is beneficial as it prevents the MOOEA from prematurely converging to just a few solutions having many copies, as can be the case, and although it is limited to intra-generational checking, as each preceding generation also has zero duplicates, it seems to perform well, even though it does not prevent previously rejected solutions from re-appearing subsequently (as they might if eliminated due only to breaking crowding limits). Nevertheless, not having to save every solution ever produced can be a significant memory saving, especially for long runs having a great many generations.

This self-adaptive GA has been shown to work well on a benchmark 2D aerodynamic design problem, as a real-world engineering example, and to provide better convergence than MOTS and in some cases NSGA-II while not being worse than either overall. It does not provide as good a density of solutions in the Pareto-optimal front as MOTS always. This can be viewed as a trade-off between being better at exploring the search space widely but doing less well at exploiting solutions found locally.

6 Future Work

It is considered desirable to enable the GA to switch between modes centred on either convergence or distribution, since these have been shown generally to be mutually exclusive. Like most generational GAs, this GA is distribution-centric as driven by its ranking algorithm, but allowing it to adopt a convergence-centric mode when convergence halts, by only allowing new solution points that dominate at least one of the existing points to enter the next generation, should improve convergence performance. Any optimisation algorithm that exhibits these characteristics is expected to be favourable for the exploration and exploitation of such complex engineering design problems. DE algorithms, in which a new solution vector only enters the population if it is better than the parent, can be only considered as convergent-centric.

Acknowledgements This work was performed with the help of a grant from the EPSRC [15] and as such is required to state that the data and methods used are held on secure media within the Power and Propulsion Group of the School of Engineering at Cranfield University (contact Professor Mark Savill).

References

1. Abott, I., von Doenhoff, A.: Theory of wing sections: including a summary of airfoil data. Dover, New York (1959)
2. Aerospaceweb: Angle of attack and pitch angle (2012). http://www.aerospaceweb.org/question/aerodynamics/q0165.shtml
3. Avidan, T., Avidan, S.: Parallax - a data mining tool based on parallel coordinates. Comput. Stat. **14**(1), 79–89 (1999). http://dx.doi.org/10.1007/PL00022707
4. Bäck, T.: Self-adaptation in genetic algorithms. In: Proceedings of the First European Conference on Artificial Life, pp. 263–271. MIT Press, Cambridge (1992)
5. Bäck, T., Eiben, A., van der Vaart, N.: An empirical study on GAs "without parameters". In: Schoenauer, M., Deb, K., Rudolph, G., Yao, X., Lutton, E., Merelo, J., Schwefel, H.P. (eds.) Parallel Problem Solving from Nature PPSN VI, vol. 1917, pp. 315–324. Springer, Heidelberg (2000)
6. Bleuler, S., Laumanns, M., Thiele, L., Zitzler, E.: Pisa - a platform and programming language independent interface for search algorithms. Evolutionary Multi-Criterion Optimization (EMO 2003), pp. 494–508. Springer, Heidelberg (2003)
7. Coello, C.A.C.: Evolutionary multi-objective optimization: a historical view of the field. IEEE Comput. Intell. Mag. **1**(1), 28–36 (2006)
8. Deb, K.: Multi-Objective Optimization using Evolutionary Algorithms. Wiley, Chichester (2001). ID: 2
9. Deb, K., Agrawal, R.B.: Simulated binary crossover for continuous search space. Complex Syst. **9**, 115–148 (1995)
10. Deb, K., Agrawal, S., Pratap, A., Meyarivan, T.: A fast elitist non-dominated sorting genetic algorithm for multi-objective optimization: NSGA-II. In: Schoenauer, M., Deb, K., Rudolph, G., Yao, X., Lutton, E., Merelo, J., Schwefel, H.P. (eds.) Parallel Problem Solving from Nature PPSN VI, vol. 1917, pp. 849–858. Springer, Heidelberg (2000)
11. Deb, K., Goyal, M.: A combined genetic adaptive search (geneas) for engineering design. Comput. Sci. Inform. **26**, 30–45 (1996)
12. Drela, M.: Xfoil - an analysis and design system for low reynolds number airfoils. In: Low Reynolds Number Aerodynamics Conference, pp. 1–12. Notre Dame, Germany (1989)
13. Eiben, A., Schut, M., de Wilde, A.: Is self-adaptation of selection pressure and population size possible? A case study. In: Runarsson, T., Beyer, H.G., Burke, E., Merelo-Guervs, L., Whitley, L., Yao, X. (eds.) Parallel Problem Solving from Nature - PPSN IX, vol. 4193, pp. 900–909. Springer, Heidelberg (2006)
14. Eiben, A.E., Schut, M.C., Wilde, A.R.D.: Boosting genetic algorithms with self-adaptive selection. In: Proceedings of the IEEE Congress on Evolutionary Computation, pp. 1584–1589 (2006)
15. EPSRC - Engineering and Physical Sciences Research Council: (2014). http://www.epsrc.ac.uk
16. Fleming, P.J., Purshouse, R.C., Lygoe, R.J.: Many-objective optimization: an engineering design perspective. In: EMO'05, LNCS 3410, pp. 14–32. Springer, Heidelberg (2005)
17. Fonseca, C.M., Fleming, P.J.: Multiobjective optimization and multiple constraint handling with evolutionary algorithms - part I: a unified formulation. IEEE Trans. Syst. Man Cybern. Part A: Syst. Humans **28**(1), 26 (1998)

18. Goldberg, D.E.: Genetic Algorithms in Search, Optimization and Machine Learning. Addison-Wesley, Massachusetts (1989)
19. Ho, C.W., Lee, K.H., Leung, K.S.: A genetic algorithm based on mutation and crossover with adaptive probabilities. In: Proceedings of the 1999 Congress on Evolutionary Computation, 1999. CEC 99, vol. 1, p. 775 (1999)
20. Holland, J.H.: Adaptation in Natural and Artificial Systems: An Introductory Analysis with Applications to Biology, Control and Artificial Intelligence, 2nd edn. MIT Press, Massachusetts (1992)
21. Inselberg, A.: Parallel Coordinates: Visual Multidimensional Geometry and Its Applications. Springer, New York (2009)
22. Jaeggi, D.M., Parks, G.T., Kipouros, T., Clarkson, P.J.: The development of a multi-objective tabu search algorithm for continuous optimisation problems. Eur. J. Oper. Res. **185**(3), 1192–1212 (2008). ID: 3
23. Jones, D.F., Mirrazavi, S.K., Tamiz, M.: Multi-objective meta-heuristics: an overview of the current state-of-the-art. Eur. J. Oper. Res. **137**(1), 1–9 (2002)
24. Jong, K.A.D.: An analysis of the behavior of a class of genetic adaptive systems. (1975). AAI7609381
25. Kipouros, T., Inselberg, A., Parks, G., Savill, A.M.: Parallel coordinates in computational engineering design AIAA 2013-1750. In: 54th AIAA/ASME/ASCE/AHS/ASC Structures, Structural Dynamics, and Materials Conference, Structures, Structural Dynamics, and Materials and Co-located Conferences. American Institute of Aeronautics and Astronautics, Boston, Massachusetts (2013). Doi:10.2514/6.2013-1750
26. Kipouros, T., Mleczko, M., Savill, M.: Use of parallel coordinates for post-analyses of multi-objective aerodynamic design optimisation in turbomachinery. AIAA-2008-2138. In: 4th AIAA Multi-Disciplinary Design Optimization Specialist Conference, Structures, Structural Dynamics, and Materials and Co-located Conferences. American Institute of Aeronautics and Astronautics, Schaumburg, Illinois (2008)
27. Kipouros, T., Peachey, T., Abramson, D., Savill, M.: Enhancing and developing the practical optimisation capabilities and intelligence of automatic design software AIAA 2012-1677. In: 8th AIAA Multidisciplinary Design Optimization Specialist Conference (MDO). American Institute of Aeronautics and Astronautics (2012)
28. Kukkonen, S., Deb, K.: Improved pruning of non-dominated solutions based on crowding distance for bi-objective optimization problems. In: IEEE Congress on Evolutionary Computation, 2006. CEC 2006, pp. 1179–1186 (2006). ID: 1
29. Li, M., Cai, Z., Sun, G.: An adaptive genetic algorithm with diversity-guided mutation and its global convergence property. J. Cent. South Univ. Technol. **11**(3), 323–327 (2004). http://dx.doi.org/10.1007/s11771-004-0066-6
30. Oliver, J.M., Kipouros, T., Savill, A.M.: A self-adaptive genetic algorithm applied to multi-objective optimization of an airfoil. In: Emmerich, M., Deutz, A., Schuetze, O., Bäck, T., Tantar, E., Tantar, A.A., del Moral, P., Legrand, P., Bouvry, P., Coello, C.A. (eds.) EVOLVE - A Bridge between Probability, Set Oriented Numerics, and Evolutionary Computation IV. Advances in Intelligent Systems and Computing, vol. 227, pp. 261–276. Springer International Publishing, Cham (2013). http://dx.doi.org/10.1007/978-3-319-01128-8_17
31. Sareni, B., Regnier, J., Roboam, X.: Recombination and self-adaptation in multi-objective genetic algorithms. In: Liardet, P., Collet, P., Fonlupt, C., Lutton, E., Schoenauer, M. (eds.) Artificial Evolution, vol. 2936, pp. 115–126. Springer, Heidelberg (2004)
32. Schaffer, J.D.: Some experiments in machine learning using vector evaluated genetic algorithms. Ph.D. thesis, Department of Electrical Engineering, Vanderbilt University, Nashville, TN (1984)
33. Sederberg, T.W., Parry, S.R.: Free-form deformation of solid geometric models. SIGGRAPH Comput. Graph. **20**(4), 151–160 (1986)
34. Selig, M.: Naca 0012 airfoils (2014). http://aerospace.illinois.edu/m-selig/ads/afplots/n0012.gif

35. Siirtola, H.: Direct manipulation of parallel coordinates. In: Proceedings of IEEE International Conference on Information Visualization, 2000, pp. 373–378 (2000). ID: 1
36. Siirtola, H., Raiha, K.J.: Interacting with parallel coordinates. Interact. with Comput. **18**(6), 1278–1309 (2006)
37. Smith, J.E., Fogarty, T.C.: Self adaptation of mutation rates in a steady state genetic algorithm. In: Proceedings of the 1996 IEEE Conference on Evolutionary Computation, pp. 318–323. IEEE (1996)
38. Storn, R., Price, K.: Differential evolution - a simple and efficient heuristic for global optimization over continuous spaces. J. Glob. Optim. **11**(4), 341–359 (1997)
39. Tan, K.C., Chiam, S.C., Mamun, A.A., Goh, C.K.: Balancing exploration and exploitation with adaptive variation for evolutionary multi-objective optimization. Eur. J. Oper. Res. **197**(2), 701–713 (2009)
40. Tan, K.C., Goh, C.K., Yang, Y.J., Lee, T.H.: Evolving better population distribution and exploration in evolutionary multi-objective optimization. Eur. J. Oper. Res. **171**(2), 463–495 (2006)
41. Tusar, T., Filipic, B.: Differential evolution versus genetic algorithms in multiobjective optimization. In: Proceedings of the 4th International Conference on Evolutionary Multi-Criterion Optimization. EMO'07, pp. 257–271. Springer, Heidelberg (2007)
42. Zhang, J., Sanderson, A.C.: Jade: Self-adaptive differential evolution with fast and reliable convergence performance. In: IEEE Congress on Evolutionary Computation. CEC 2007, pp. 2251–2258. (2007) ID: 1
43. Zhang, J., Sanderson, A.C.: Self-adaptive multi-objective differential evolution with direction information provided by archived inferior solutions. In: IEEE Congress on Evolutionary Computation, CEC 2008. (IEEE World Congress on Computational Intelligence), pp. 2801–2810. (2008) ID: 1
44. Zitzler, E., Thiele, L.: Multiobjective optimization using evolutionary algorithms - a comparative case study. In: Parallel Problem Solving From Nature - PPSN V, pp. 292–301. Springer, Heidelberg (1998)
45. Zitzler, E., Thiele, L.: Multiobjective evolutionary algorithms: a comparative case study and the strength pareto approach. IEEE Transactions on Evolutionary Computation, **3**(4), 257–271 (1999). ID: 1
46. Zitzler, E., Thiele, L., Laumanns, M., Fonseca, C.M., Grunert da Fonseca, V.: Performance assessment of multiobjective optimizers: an analysis and review. IEEE Trans. Evol. Comput. **7**(2), 117–132 (2003)

Evidence-Based Multi-disciplinary Robust Optimization for Mars Microentry Probe Design

Liqiang Hou, Yuanli Cai and Jisheng Li

Abstract Atmospheric pressure on Mars is approximately 1 % of that on Earth and varies about 15 % during the year due to condensation and sublimation of its primarily CO_2 atmosphere. Impacts of the uncertainties during the entry are difficult to be modeled. The situation becomes more complex when uncertainties are from different disciplines. In this work, a robust multi-disciplinary optimization method for Mars microentry probe design under epistemic uncertainties is presented. Objectives of the evidence-based robust design are set to minimize the interior temperature of thermal protection systems (TPS) and maximize its belief value under uncertainties. A population-based multi-objective estimation of distribution algorithm (MOEDA) is designed for searching the robust Pareto set. Candidate solutions are adaptively clustered into groups. In each group, principal component analysis (PCA) technique is performed to estimate population distribution, sample and reproduce individuals. Non-dominated individuals are sorted and selected through the NSGA-II-like selection procedure. Adaptive sampling and binary branching techniques are employed for computing the evidence belief functions. PCA dimensionality reduction technique is implemented for identifying and removing uncertain boxes with little contribution of the beliefs. With variable fidelity model management, analytical aerodynamic model is used first to initialize the optimization searching direction. Artificial neural network (ANN) surrogate model is used for reducing the computational cost. When the optimization goes close to the optima, more data from the high accuracy model are put into the aerodynamic database, making the optimization procedure converge on optima quickly while keeping high-level accuracy.

L. Hou (✉)
State Key Laboratory of Astronautic Dynamics, Xi'an Satellite Control Center,
No. 462, Xianing East Road, Xi'an 710043, China
e-mail: houliqiang2008@139.com

Y. Cai · J. Li
School of Electronic and Information Engineering, Xi'an Jiaotong University,
No. 28, Xianing West Road, Xi'an 710049, China

© Springer International Publishing AG 2017
M. Emmerich et al. (eds.), *EVOLVE – A Bridge Between Probability,
Set Oriented Numerics and Evolutionary Computation VII*,
Studies in Computational Intelligence 662, DOI 10.1007/978-3-319-49325-1_7

1   Introduction

When designing Mars microentry probe, challenges for the designers are complicated by many factors. Atmospheric pressure on Mars is approximately 1 % of that on Earth and varies about $\pm 15\%$ during the year due to condensation and sublimation of its primarily CO_2 atmosphere. Temperature on the Mars surface might be cold enough that carbon dioxide freezes during the winter and "snows" onto the polar cap. Therefore, uncertain impacts of these factors should not be neglected during the mission design, and a robust design optimization method taking uncertainties into account is required.

Numerous examples of multi-disciplinary design optimization (MDO) applications have been found in many areas [4, 23, 26]. Roshanian proposed an integrated approach for multi-disciplinary design of launch vehicle, using response surface method for approximation of the propulsion model [25]. Huang proposed a multi-objective Pareto concurrent subspace optimization method for multi-disciplinary design [15]. In [11], interval-based methods were used in the multi-objective optimization (MOO). To reduce computational cost, some researchers introduced variable fidelity model management into MDO for the reentry vehicle and air-craft design [10, 21]. Taking into account the uncertainty impacts, Mueller and Larsson proposed a robust optimization method for collision avoidance maneuver planning [20]. Lantoine proposed a hybrid differential dynamic programming algorithm for robust low-thrust optimization [17]. Vasile and Croisard proposed a robust mission design method with Evidence Theory [6, 27].

In real-life engineering design problems, particularly some problems in the preliminary design phase, it is generally desirable to investigate uncertainty impacts in the design optimization. An appropriate representation of the uncertainty in analysis outcomes is therefore an essential part of complete analysis. For aleatory uncertainties, whose probability density functions are supposed to be known, probability-based analysis tools can be used for computing the confidence level of the solutions, and the traditional optimization tools can be used for searching the optimal solutions with probability-based reliable constraints. However, the knowledge of the probability distribution is not always available for the uncertainties. For epistemic uncertainties due to lack of knowledge, the uncertainties are expressed by means of intervals based on experts' opinion or rare experimental data, and the classic probability theory-based analysis tools cannot be used. Therefore, some researchers proposed evidence-based robust design optimization methods for the design optimization under epistemic uncertainties [6, 27].

With Evidence Theory (Dempster-Shafer's theory), both aleatory and epistemic uncertainties, coming from a poor or incomplete knowledge of the model parameters, can be correctly modeled [1]. The values of uncertain parameters are expressed by means of intervals with associated probabilities. In particular, the value of belief expresses the lower probability that the selected design point remains optimal (and feasible) even under uncertainties.

In this work, a population-based robust optimization (RO) method for Mars microentry probe design is proposed. The optimization objective is to minimize the interior temperature of TPS, while at the same time to maximize its belief value. Candidate solutions are adaptively grouped with affinity propagation method. No predefined number of the clusters is required. In each cluster, principal component analysis (PCA) is performed for estimating population distribution, sampling, and reproducing individuals. Non-dominated individuals are sorted and selected through the NSGA-II-like selection procedure. Multi-fidelity aerodynamic models are integrated into the robust optimization procedure. Variable fidelity model management is conducted through artificial neural network (ANN) surrogate model. When the optimization goes close to the optima, more data from the high accuracy model are put into the aerodynamic database, making the optimization procedure converge on optima quickly while keeping high-level accuracy.

The rest of the paper is organized as follows: Sect. 2 describes dynamic models involved in this paper, including the entry dynamic, aerodynamic, and thermal dynamic models. Section 3 investigates impacts of the epistemic uncertainties and corresponding computation of evidence levels. The multi-objective robust optimization method is presented in Sect. 4. Numerical simulation of the microprobe (no more than 0.8 m in diameter) design under uncertainty is given in Sect. 5, and conclusions are presented in Sect. 6.

2 Dynamic Models

2.1 Entry Dynamic Equations

With state variable $[r, \lambda, \phi, v, \theta_p, \xi]^T$, the entry dynamic equations in planet-centered frame can be written as [24, 29] (Fig. 1)

$$\dot{r} = v \sin \theta_p$$

$$\dot{\lambda} = v \frac{\cos \theta_p \cos \xi}{r \cos \phi}$$

$$\dot{\phi} = \frac{v \cos \theta_p \sin \xi}{r}$$

$$\dot{v} = -\frac{D}{m} - g \sin \theta_p$$

$$\dot{\theta}_p = \frac{L}{mv} \cos \gamma_v - \left(\frac{g}{v} - \frac{v}{r} \right) \cos \theta_p$$

$$\dot{\xi} = \frac{L}{mv \cos \theta_p} \sin \gamma_v - \frac{v}{r} \cos \theta_p \cos \xi \tan \phi \qquad (1)$$

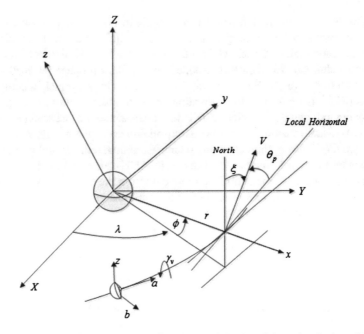

Fig. 1 Illustration of point mass entry trajectory equations: λ and ϕ are longitude and latitude, respectively; θ_p denotes flight path angle, and ξ is velocity azimuth angle

where r, λ, and ϕ are the distance from the center of the planet to the vehicle, longitude, and latitude, respectively; θ_p denotes flight path angle, and ξ denotes velocity azimuth angle. Drag D and lift L in the Eq. 1 are given by

$$D = \frac{1}{2}\rho(h)SC_d v^2$$
$$L = \frac{1}{2}\rho(h)SC_l v^2 \qquad (2)$$

Given the initial condition, state variables at each point of the trajectory can be obtained by integrating Eq. 1. The aeroshell of the probe is defined by a set of geometric parameters (see Fig. 2). The parameters consist of radius of nose R_n, diameter R_b, and semi-apex angle θ. The leeward side of the body is a semisphere.

With modified Newtonian theory, drag coefficient and lift coefficient can be obtained [3]

$$C_d = C_d(C_{pt2}, \alpha, \gamma, R_n, \theta, R_b), \quad C_l = C_l(C_{pt2}, \alpha, \gamma, R_n, \theta, R_b) \qquad (3)$$

where α is angle of attack (AoA), γ is specific heat rate, and pressure coefficient C_{pt2} can be computed with

Fig. 2 Geometric definition of the micro Mars probe.

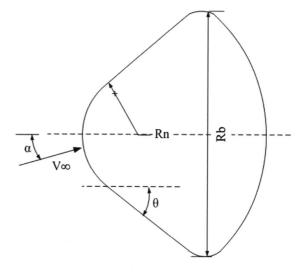

$$C_{pt2} = \frac{2}{\gamma}\left(\frac{\gamma+1}{2}\right)\frac{\gamma}{\gamma-1}\left(\frac{\gamma+1}{2\gamma-\frac{\gamma-1}{M_\infty^2}}\right)^{\frac{1}{\gamma-1}} - \frac{2}{\gamma M_\infty^2} \tag{4}$$

where M_∞ is Mach number.

Suppose that $M_\infty \gg 1.0$, lift coefficient and drag coefficient can be obtained with Eqs. 3 and 4. More accurate data taking into account all effects can be from numerical computational fluid dynamics (CFD) model. In this work, the commercial CFD software Numeca® is used to perform high-accuracy computation of Reynolds-averaged Navier–Stokes equations. However, the costly computation makes it impractical to rely exclusively on the high-fidelity CFD model for the design optimization.

2.2 Heat Flux and Heat Load

The thermal protection system (TPS) is composed of SLA-561V, a material widely used in space engineering for thermal protection, and has been used as the primary TPS material on all the sphere-cone Mars entry vehicles sent by NASA. Equation for computing heating flux on the TPS surface is given by [3, 19]

$$\dot{q} = 1.89^{-8}\sqrt{\rho/R_n}v^3 \tag{5}$$

Suppose that the heat transfer occurs only in one direction deep into the TPS layer, the one-dimensional heating transfer equation can then given by [2, 13]

$$\frac{\partial}{\partial x}\left(k_c \frac{\partial T}{\partial x}\right) + \dot{m}_p c_p \frac{\partial T}{\partial x} = \rho_c c_p \frac{\partial T}{\partial t}$$

$$\frac{\partial}{\partial x}\left(k_v \frac{\partial T}{\partial x}\right) = \rho_v c_p \frac{\partial T}{\partial t} \tag{6}$$

with boundary conditions under radiation equilibrium condition:

$$\dot{q} = \varepsilon \sigma T_w^4 + k\frac{dT}{dx} \quad (x = 0)$$

$$k\frac{dT}{dx} = \varepsilon \sigma \mathrm{T}_{in}{}^4 \quad (x = L_{TPS}) \tag{7}$$

where c_p is heat capacity, $\dot{m}$ is surface recession rate of TPS materials, ρ_c, and ρ_v denotes the density of char material and virgin material; k_c and k_v are thermal conductivities of char material and virgin material, respectively; T_w and T_{in} denote temperatures at outer and inner surfaces. Material properties of SLA-561V are taken from TPSX website [14]. Integrating Eqs. 5–7, temperature distribution of TPS can be obtained with finite-difference method.

Neglecting the effects of decomposition, and pyrolysis gas flow, with the semi-infinite solid approximation, closed-form analytical solutions to the in-depth heat transfer equation can be obtained. The simplified relation of TPS surface temperature T_s and heat flux $\dot{q}$ is

$$\dot{q} = \varepsilon \sigma T_s^4 \tag{8}$$

where ε is emissivity factor, and $\sigma = 5.6703e - 8$ is the Stefan-Boltzmann constant, respectively. The temperature at a depth x within the solid at time t is given by

$$T(x, t) = erf\left(\frac{x - \dot{x}t}{2\sqrt{\alpha t}}\right)(T_0 - T_s) + T_s \tag{9}$$

where erf is the gaussian error function, $\dot{x}$ is TPS surface recession rate, and T_0 is the initial temperature, respectively. The thermal diffusivity α in Eq. 9 is given by

$$\alpha = \frac{k}{\rho c_p} \tag{10}$$

where k is thermal conductivity, c_p is specific heat, and ρ is density, respectively. The equations above can be used as simplified model for conceptual design in preliminary design phase.

3 Uncertain Impacts and Evidence Theory Modeling

3.1 Evidence Theory-Based Uncertainty Modeling

In this section, an Evidence Theory (ET)-based uncertainty modeling technique is presented. In this method, as opposed to a single value of probability, bounds for uncertainty quantification are used instead. Propagation of the information is through basic probability assignment (BPA) [1, 6, 27]. The total degree of belief in a proposition A is expressed within a bound $[Bel(A), Pl(A)]$ lying in the unit interval $[0, 1]$. $Bel(A)$ is obtained by accumulation of BPAs of the propositions that imply proposition A, whereas $Pl(A)$ is plausibility calculated by adding the BPAs of propositions that imply or could imply the proposition A.

In Evidence Theory, the belief of an uncertain parameter $u \in [a, b]$ is an elementary proposition. The level of confidence an expert has on an elementary proposition is quantified using the basic probability assignment (BPA). The BPA satisfies the following rules:

$$m(E) \geq 0, \forall E \in \mathbf{U}$$
$$m(\emptyset) = 0$$
$$\sum_{E \in \mathbf{U}} m(E) = 1 \tag{11}$$

where $m(E)$ is BPA of the elementary proposition E, and $\mathbf{U}$ is a set that contains all possible E and unions of E.

A focal element (FE) is an element of $\mathbf{U}$ that has a non-zero BPA. The Belief (Bel) and Plausibility (Pl) functions of proposition A are [1, 6]:

$$Bel(A) = \sum_{FE \subset A, \; FE \subset \mathbf{U}} m(FE) \tag{12}$$

$$Pl(A) = \sum_{FE \cap A \neq \emptyset, \; FE \subset \mathbf{U}} m(FE) \tag{13}$$

For BPAs of more than one uncertain parameters (e.g., u_1 and u_2, and corresponding intervals they are located in, $[a_1, b_1]$ and $[a_2, b_2]$), as Eq. 14 shown, the BPA of given Cartesian product is product of the BPA of each interval, i.e.,

$$m((u_1, u_2) \in [a_1, b_1] \times [a_2, b_2]) = m(u_1 \in [a_1, b_1]) \times m(u_2 \in [a_2, b_2]) \tag{14}$$

Using Evidence Theory for robust engineering design was proposed in 2002 by Oberkampf et al. [22] and was recently applied to robust design of space systems [1, 18] and space trajectory design [6]. Suppose the uncertain parameters and design variables are $\mathbf{u} = [u_1, u_2, ..., u_m] \in \mathbf{U}$ and $\mathbf{d} = [d_1, d_2, ..., d_n] \in \mathbf{D}$, respectively, where

$\mathbf{U} \in \mathbf{R}^m$ and $\mathbf{D} \in \mathbf{R}^n$ are uncertain space and design space, respectively, the robust optimization can be formulated as

$$\max_{v \in \mathbb{R}, \mathbf{x} \in \mathbf{D}} Bel(f(\mathbf{d}, \mathbf{u}) < v)$$

$$\min_{v \in \mathbb{R}, \mathbf{x} \in \mathbf{D}} v \qquad (15)$$

where $v \in \mathbb{R}$ is the threshold to be minimized, and $f(\mathbf{d}, \mathbf{u})$ is the main objective function. Based on this idea, M. Vasile et al. proposed three approaches to solve the OUU (optimization under uncertainty) problem: evolutionary multi-objective approach, step technique, and clustering approximation method [6]. These approaches are applied to BepiColombo preliminary mission design, and minimize wet mass of the spacecraft taking into account uncertainties. In the following sections, a population-based multi-objective optimization method is proposed for the robust design optimization.

3.2 Approximation of Belief and Plausibility

Given proposition $A = \{\mathbf{u} \in \mathbf{U} | f(\mathbf{d}, \mathbf{u}) \le v\}$, and subsets of Θ, belief and plausibility of proposition A can be computed through

$$Bel(A) = \sum_{\forall \theta_i \subseteq A} m(\theta_i), \qquad Pl(A) = \sum_{\forall \theta_i \cap A \neq 0} m(\theta_i) \qquad (16)$$

where $m(\theta)$ is the BPA associated to the subset θ_i, and $\mathbf{U}$ is the uncertain space.

In this work, a binary tree evolutionary algorithm is performed for computing the belief and plausibility values [28]. In this algorithm, with uncertainty space $\mathbf{U}$ and threshold v, an iterative partitioning and pruning procedure is implemented for identifying those focal elements contributing to the plausibility and belief of v. Algorithm 1 shows how the binary tree evolutionary algorithm works with the proposition $A = \{\mathbf{u} \in \mathbf{U}_i | f(\bar{\mathbf{x}}, \mathbf{u}) \le v_i\}$, where $\bar{\mathbf{x}}$ is design vector, v_i is threshold, and $\mathbf{U}_i$ is the uncertain box.

3.3 Dimensionality Reduction of Uncertainties

With Evidence Theory, computational cost for Bel and Pl grows exponentially with the number of the dimensions. In this work, a double-level strategy is implemented for obtaining the approximation of the Belief and Plausibility cumulative function. First, a principle component analysis (PCA)-based algorithm is implemented for dimensionality reduction, leaving those focal elements contributing most for Bel and Pl values.

Algorithm 1 Binary tree evolutionary algorithm

1: Given design parameter $\bar{\mathbf{x}}$, and uncertain box $\mathbf{U}_i$
2: **for** each threshold v_i **do**
3:   $iter = 0$
4:   initialize the uncertain measure values with $Bel(v) = 0$ and $Pl(v) = 0$
5:   partition hypercube $\mathbf{U}_i$ into $\bar{\mathbf{U}}^l$ and $\bar{\mathbf{U}}^r$ along the longest edge
6:   **for** each sub-box $\bar{\mathbf{U}}^r, \bar{\mathbf{U}}^l$ **do**
7:     compute $f_{min}^j = \min\limits_{\mathbf{u} \in \bar{\mathbf{U}}^j} f(\bar{\mathbf{x}}, \mathbf{u})$ and $f_{max}^j = \max\limits_{\mathbf{u} \in \bar{\mathbf{U}}^j} f(\bar{\mathbf{x}}, \mathbf{u})$, where $j = l, r$
8:     **if** $f_{min}^j > v$ **then**
9:       $pl(-A) = pl(-A) + BPA(\bar{U}^j)$, remove the box $\bar{U}^j$
10:     **else**
11:       **if** $f_{max}^j < v$ **then**
12:         $Bel(A) = Bel(A) + BPA(\bar{\mathbf{U}}^j)$, $Pl(A) = Pl(A) + BPA(\bar{\mathbf{U}}^j)$
13:       **end if**
14:     **end if**
15:   **end for**
16:   $iter = iter + 1$
17:   **if** $iter < n_{max}$ and $BPA(\bar{\mathbf{U}}^j) > \varepsilon$ **then**
18:     **go to** 5
19:   **end if**
20: **end for**

In this analysis, given design parameters, a series of numerical computations with uncertainty impacts are performed. This set is chosen to represent the impacts of uncertainties. A Latin hypecube sampling of uncertainties are performed and is stored in a $M \times N$ matrix, where M is the number of experiments and N is the size of variables $\Psi = f(\mathbf{d}, \mathbf{u})$

$$\Psi = \begin{pmatrix} \Psi_1^{(1)} & \Psi_1^{(2)} & \cdots & \Psi_1^{(M)} \\ \Psi_2^{(1)} & \Psi_2^{(2)} & \cdots & \Psi_2^{(M)} \\ \vdots & \vdots & \vdots & \vdots \\ \Psi_N^{(1)} & \Psi_N^{(1)} & \cdots & \Psi_N^{(M)} \end{pmatrix} \qquad (17)$$

with numerical samples Ψ, a deviation matrix computed with respect to the mean vector $\bar{\Psi}$ can be obtained as

$$\tilde{\Psi} = \Psi - \bar{\Psi} \qquad (18)$$

With covariance matrix $\mathbf{C} = \tilde{\Psi} \cdot \tilde{\Psi}^T$, a linear basis can be obtained by extracting eigenvectors $\mathbf{U}_i$ from $\mathbf{C}$. After ranking the eigenvectors in descending order of their corresponding eigenvalues, relative contributions of each uncertainty can be estimated. Equation 19 shows the criterion used for dimensionality reduction

$$e = 1 - \frac{\sum_{i=1}^{m} \lambda_i^2}{\sum_{i=1}^{M} \lambda_i^2} \qquad (19)$$

BPAs of the discarded boxes are added up to those boxes remained such that sum of BPAs over the whole uncertainty space holds 1.0.

After dimensionality reduction, to further the approximation, boxes with BPA lower than the specified value are removed for their little contributions of *Bel* and *Pl*. Sort the boxes with respect to their BPAs, remove those boxes whose BPA values are less than, e.g., 0.01, and transfer the BPAs to the remainder boxes related to uncertain parameters. EBT technique can then be implemented for computing *Bel* and *Pl* values.

3.4 Adaptive Sampling for Belief Function Computation

Note that in the computation of $Pl(A)$ and $Pl(-A)$ in EBT, when exploring solution space in $\bar{U}^l$ and $\bar{U}^r$, it does not always require exact minimum (maximum) value of the cost function. One can consider some other approximate values instead. This will help reduce computational cost, but may introduce error during the computation. This is implemented by an adaptive sampling and selecting strategy.

In the evidence-based robust optimization, the computational cost mainly comes from those programs searching minimum and maximum values of the cost function. The cost is particularly expensive for reentry vehicle design in which large number of uncertainties and design parameters are involved. In this study, instead of conducting the minimum searching programs such as fmincon function of MATLAB, an adaptive sampling and searching technique is used. Samples are selected in the uncertainty box with uniform distribution. Those values of maximum and minimum magnitude are selected and put into the archive for performing the computation of belief functions. For each individual box, sample size is set to be proportional to the product of BPA and longest distance of the box. This is based on the assumption that minimum and maximum values are uniformly distributed and varies monotonously with respect to the uncertainties.Therefore, one can use the samples instead of searching the real maximum values at very high expense of computational cost.

4 Multi-fidelity Robust Optimization

4.1 Multi-objective Optimization with Adaptive Clustering Density Estimator

In this section, a population-based multi-objective estimation of distribution algorithm (MOEDA) is proposed for conducting the multi-objective optimization (MOO). Individuals are adaptively clustered into groups. In each group, local principle component analysis is used for modeling distribution and reproduction of individuals. The method is organized as follows:

1. Clustering Using Affinity Propagation

There are several ways to cluster a data set. The popular k-centers clustering technique begins with a set of randomly selected exemplars and iteratively refines this set so as to decrease the sum of squared errors. Another similar method is to use local PCA for partitioning population [30]. Both algorithms have the number of clusters predefined before data partitioning is performed.

In this paper, affinity propagation clustering technique is used instead [12]. In this algorithm, affinity propagation takes as input a real number $s(k, k)$, "preference," for each data point. Data points with larger values of $s(k, k)$ are more likely to be chosen as exemplar. Real-valued messages are exchanged between data points until high-quality centers and corresponding clusters gradually emerges. The "responsibility," $r(i, k)$, sent from data point i to candidate exemplar point k, reflects the accumulated evidence for how well-suited point k is to serve as exemplar for point i. The "availability," $a(i, k)$, sent from candidate exemplar point k to point i reflects the accumulated evidenced for how appropriate it would be for point i to choose k as its exemplar. Figure 3 shows how the messages of "responsibilities" and "availabilities" work.

In affinity propagation clustering method, the number of clusters need not be specified beforehand. Instead, the appropriate number of clusters will be determined from the message passing and depends on the input exemplar preference. The initial preference is set to be the median of the input similarities, a negative squared error. Real-valued messages are exchanged between data points until high-quality centers and corresponding clusters gradually emerges.

2. Modeling Using Principal Component

In each cluster, principle component technique is employed for modeling individuals [30]. For each cluster S^j, boundaries of the subset are given by

$$d_i^j = \min_{x \in S^j} (x - \bar{x}_j)^T U_i^j, \quad b_i^j = \max_{x \in S^j} (x - \bar{x}_j)^T U_i^j \qquad (20)$$

where the principal component U^j is computed as a unit eigenvector associated with the sample data covariance matrix C of the points in S^j

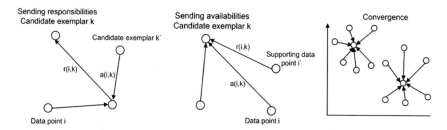

Fig. 3 Affinity propagation clustering: individuals are clustered into groups with messages of "responsibilities" and "availability" between individuals

$$C = \frac{1}{|S| - 1} \sum_{x \in S} (x - \bar{x})(x - \bar{x})^T \tag{21}$$

where $|S|$ is cardinality of S^j. With the ith largest eigenvalue λ_i^j of C in S^j, we have

$$\sigma_j = \frac{1}{n - m + 1} \sum_{i=m}^{n} \lambda_i^j \tag{22}$$

where m is the number of objectives and n is the dimension of the vector x.

To make the clusters more approximate the Pareto set (PS), and avoid missing real solutions, extensions on both sides of boundary are made to generate new sampling subspace Ψ^j. The percentage the extension made is 25 % of either side:

$$\Psi^j = \left\{ x \in R^n \,\middle|\, x = \bar{x}^j + \sum_{i=1}^{m-1} \alpha_i U_i^j, \, i = 1, \dots, m - 1 \right\} \tag{23}$$

with

$$\alpha_i^j - 0.25(b_i^j - a_i^j) \leq \alpha_i \leq b_i^j + 0.25(b_i^j - a_i^j) \tag{24}$$

3. Reproduction by Sampling

Generate τ samples with the probability proportional to its size for subspace Ψ^k:

$$P(\tau = k) = \frac{vol(\Psi^k)}{\sum_{j=1}^{k} vol(\Psi^j)} \tag{25}$$

Uniformly randomly generate a point x' from Ψ^τ. Generate N new individuals $x = x' + \varepsilon$, where ε is a noise vector from $N(0, \sigma_\tau I)$.

4. Selection

The selection procedure is based on the non-dominated sorting similar to NSGA-II [7]. Crowding distance of the points in set is defined as the average side length to the largest m-D rectangle in the objective space. Solutions with lager crowing distance will be put into the new population.

4.2 Experimental Results of the New MOO with Multi-probability Constraints

Consider a bi-objective optimization problem with two design variables $\mathbf{x} = [x, y]$ with variance $\sigma = 0.03$ [8].

$$\min \quad f_1 = x$$
$$\min \quad f_2 = \frac{1+y}{x}$$
$$\text{subject to} \quad y + 9x - 6 \geq 0,$$
$$-y + 9x - 1 \geq 0,$$
$$0.1 \leq x \leq 1,$$
$$0 \leq y \leq 5 \tag{26}$$

The probability constraint is

$$P(g_j(x) \geq 0) \geq R_j, \ (j = 1, 2, ..., J) \tag{27}$$

where $\mathbf{x}$ are design variables, R_j is the desired reliability (within $[0, 1]$). By replacing the original probability constraint by $G_j(U^*) \geq 0$, the most probable point (MPP) can be obtained by solving the optimization problem

$$\text{Minimize} \quad G_j(U)$$
$$\text{subject to} \quad \|U\| = \beta_j^r \tag{28}$$

where β_j^r is the required reliability index computed from the required reliability R_j as

$$\beta_j^r = \Phi^{-1}(R_j) \tag{29}$$

Population size of the MOEDA is set to be 50. After 100 generations, a set of reliable solutions is obtained. Figure 4 shows the deterministic front and three reliable frontiers with reliability index β^r equal to 1.0(84.13 %), 2.0(97.725 %), and 3.0(99.875 %). As shown in the figure, the reliable trade-off frontier moves inside the feasible objective space as β^r increases. These solutions agree well with those in [8].

4.3 Multi-fidelity Model Management

In Mars entry probe design optimization, direct search methods relying exclusively on high-fidelity aerodynamic models is cost-prohibitive. Therefore, researchers proposed different model management strategies for variable-fidelity optimization [5, 16]. In this section, an ANN-based multi-fidelity strategy is designed to integrate the models with different accuracy levels into the robust optimization.

The model management works as follows: First, a Latin hype-cube sampling is carried out for different geometric configurations, with which an ANN for response surface fits is initialized by the results from both low- and high-fidelity models.

Fig. 4 Trade-off frontiers
for the two probability
constraints test problem

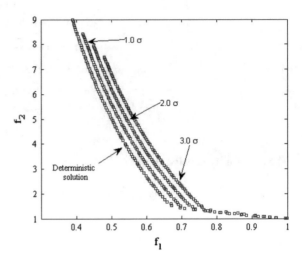

In the ANN approximator, the inputs are drag coefficient and lift coefficient from both models, and the corresponding Mach number and AoA, while the output is the difference between CFD and modified Newtonian theory. The actual lift and drag coefficients in the optimization procedure are then from the analytical model plus output of the ANN.

The ANN is updated after every few generations of optimization. Only the individuals of great value are sampled using high-fidelity model. The sample sets consist of the centroid of the cluster, individuals located on the lower and higher bounds of the sampling cluster, and other interested individuals, e.g., individuals with highest temperature in the cluster. Putting into the ANN new high-fidelity results and corresponding lower fidelity results and training the ANN, new response surface can then be established. During the optimization, more and more information from high-fidelity model is put into the surrogate model, making the surrogate model closer to the high-fidelity model, and finally after several generations, converges to the optimal values with accuracy level of the high-fidelity model.

Figure 5 shows flowchart of the evolution control and database handler.

5   Numerical Results

In this section, robust optimization of a microentry probe design is presented. The probe cross section diameter is 0.8 m, with entry mass $m = 12$ kg and TPS thickness $L_{TPS} = 1.4$ cm. The design parameters are $\mathbf{d} = [R_n, \theta]^T$. Initial values of the design parameters are given in Table 1.

Initial entry conditions are set to be same as the Spirit spacecraft and listed in Table 2, with uncertainties $\mathbf{u} = [\rho, C_d, C_l, r, \lambda, \phi, v, \theta, m]^T$ of atmospheric model and trajectory parameters. Table 3 shows their corresponding BPA structure.

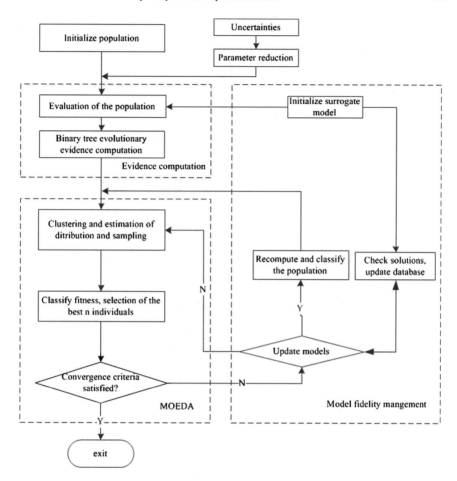

Fig. 5 Evolutionary control and database handler: relation of evidence computation, surrogate and multi-objective optimization

Table 1 Initial values of design parameters

Parameters	θ	R_n
Lower bound	35.0°	0.04 m
Upper bound	75.0°	0.15 m

Table 2 Initial entry conditions [9]

r	v	ξ	θ_p	λ	ϕ
3392.3 km	5.628 km/s	79.025 (°)	−11.495 (°)	161.776 (°)	−17.742 (°)

Table 3 BPA structure of uncertain parameters

Parameters	Lower bound	Upper bound	BPA
ρ	−10%	−5%	0.05
	−5%	0	0.25
	0	5%	0.3
	5%	10%	0.4
C_d	−10%	−5%	0.05
	−5%	0	0.25
	0	5%	0.3
	5%	10%	0.4
C_l	−10%	−5%	0.05
	−5%	0	0.25
	0	5%	0.3
	5%	10%	0.4
r	−2.0 km	−0.5 km	0.20
	−0.5 km	0	0.40
	0	0.5 km	0.40
	0.5 km	2.0 km	0.20
λ	−0.20°	−0.10°	0.20
	−0.10°	−0.05°	0.30
	−0.05°	0.10°	0.30
	0.10°	0.20°	0.20
φ	−0.20°	−0.10°	0.20
	−0.10°	−0.05°	0.30
	−0.05°	0.10°	0.30
	0.10°	0.20°	0.20
v	−50 m/s	−20 m/s	0.20
	−20 m/s	−10 m/s	0.40
	−10 m/s	10 m/s	0.30
	10 m/s	20 m/s	0.10
θ	−0.5°	−0.05°	0.40
	−0.05°	−0.01°	0.30
	−0.01°	0.01°	0.20
	0.01°	0.20°	0.10
m	−0.5 kg	−0.10 kg	0.40
	−0.10 kg	−0.05 kg	0.30
	−0.05 kg	0.10 kg	0.20
	0.10 kg	0.50 kg	0.10

ANN-based surrogate model is trained through Bayesian regularization back propagation. ANN is trained with different geometric configurations and flow conditions. The first round of training is done by putting into ANN different values of C_d and C_l from the analytical model and CFD software with respect to M_∞ and geometric configuration, 300 samples in total.

Analytical aerodynamic model is used during the first few iterations to learn the cost function. Numeca® discretizes the computational domain with multi-block structured mesh. The computational meshes consist of 18 blocks with nearly 1.2106 total nodes, changed by internal scripting based on design parameters. The flow model consist of CO_2 and N_2, of which 97% by volume is CO_2, and 3% by volume is N_2. After every 10 generations, the database is updated by inserting new results from the CFD software. Trajectory equation is integrated until altitude reaches to 10 km, or the Mach number is lower than 1.3. The objectives of the robust optimization

Fig. 6 Temperature versus Bel

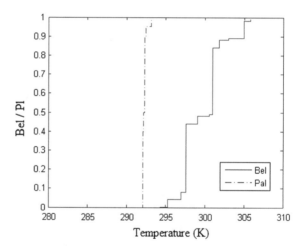

Fig. 7 Temperature versus Bel

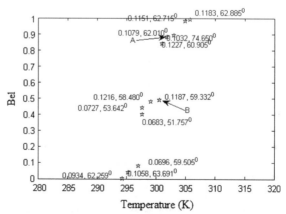

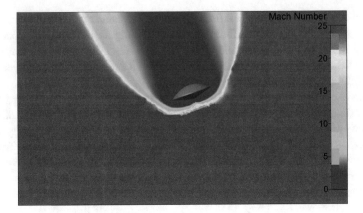

Fig. 8 Optimal solutions: individual A

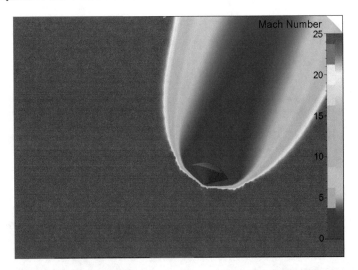

Fig. 9 Optimal solutions: individual B

are to minimize the TPS interior temperature and maximize its belief value, with a constraint that the maximum overload should be no more than 9 g.

The computation is performed on a Linux platform with 32 Core (4 × 8 3.0 GHz AMD 6220), 40 GB of memory. Figures 6 and 7 show the optimization results with a population of 60 individuals after 80 generations. Geometric parameters ($[R_n, \theta]$) of each individual on the Pareto front are listed in the figures as well (Fig. 7).

Two individuals A and B are selected from Pareto optimal front for illustrating their performance (Figs. 8 and 9). Individual A has a belief of 0.847 with highest temperature of 304.1 k, while probe B's temperature is 301.4 k with a belief of 0.487. Corresponding trajectories and TPS temperatures are shown in the figure as well (Figs. 10 and 11).

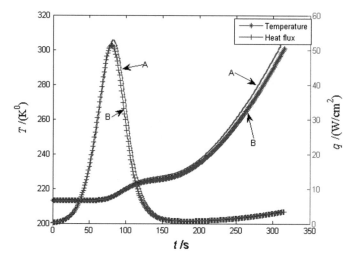

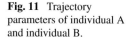

Fig. 10 Trajectory and TPS performance of individual A and individual B.

Fig. 11 Trajectory parameters of individual A and individual B.

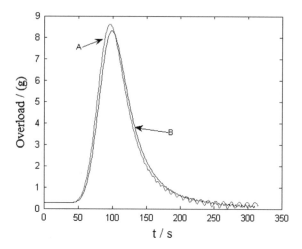

6   Conclusion

A robust multi-fidelity optimization strategy is proposed for Mars microentry probe design. Evidence Theory is implemented to model epistemic-type uncertainty impacts. Optimization results are improved further with the multi-fidelity models, reducing the computational cost while accuracy level is preserved. Adaptive clustering MOEDA is designed for conducting the multi-objective optimization, using local principle component for generating new individuals. In each cluster, local principle component techniques are employed for data modeling and data generation. A MOO with probability constraints is used to test the MOO's performance. Experimental results show that the proposed algorithm could be used in robust optimization under both the aleatory and epistemic uncertainties.

Moreover, a multi-fidelity model is incorporated into the robust multi-disciplinary optimization. ANN surrogate model is trained firstly with low-fidelity results, with high-fidelity data put into the database, the solutions converge to high-fidelity results. EBT branching and pruning technique and adaptive approximated computation are used for the computation of belief function during the robust optimization. The method in this paper provides a way for conducting optimization of the similar problems in multi-disciplinary space mission design under epistemic uncertainties.

References

1. Agarwal, H., Renaud, J.E., Preston, E.L., Padmanabhan, D.: Uncertainty quantification using evidence theory in multidisciplinary design optimization. Reliab. Eng. Syst. Saf. **85**(1), 281–294 (2004)
2. Amar, A.J., Blackwell, B.F., Edwards, J.R.: One-dimensional ablation using a full newton's method and finite control volume procedure. J. Thermophys. Heat Transf. **22**(1), 71–82 (2008)
3. Anderson, J.D.: Hypersonic and High Temperature Gas Dynamics. AIAA (2000)
4. Balesdent, M., Bérend, N., Dépincé, P.: Stagewise multidisciplinary design optimization formulation for optimal design of expendable launch vehicles. J. Spacecr. Rocket **49**(4), 720–730 (2012)
5. Cheng, Q.S., Bandler, J.W., Koziel, S.: Combining coarse and fine models for optimal design. IEEE Microw. Mag. **9**(1), 79–88 (2008)
6. Croisard, N., Vasile, M., Kemble, S., Radice, G.: Preliminary space mission design under uncertainty. Acta Astronaut. **66**(5), 654–664 (2010)
7. Deb, K., Pratap, A., Agarwal, S., Meyarivan, T.: A fast and elitist multiobjective genetic algorithm: NSGA-II. IEEE Trans. Evol. Comput. **6**(2), 182–197 (2002)
8. Deb, K., Gupta, S., Daum, D., Branke, J., Mall, A.K., Padmanabhan, D.: Reliability-based optimization using evolutionary algorithms. IEEE Trans. Evol. Comput. **13**(5), 1054–1074 (2009)
9. Desai, P.N., Knocke, P.C.: Mars exploration rovers entry, descent, and landing trajectory analysis. In: AIAA/AAS Astrodynamics Specialist Conference and Exhibit, pp. 16–19 (2004)
10. Dufresne, S., Johnson, C., Mavris, D.N.: Variable fidelity conceptual design environment for revolutionary unmanned aerial vehicles. J. Aircr. **45**(4), 1405–1418 (2008)
11. Emmerich, M., Naujoks, B.: Metamodel-assisted multiobjective optimization with implicit constraints and its application in airfoil design. In: International Conference & Advanced Course ERCOFTAC, Athens, Greece (2004)
12. Frey, B.J., Dueck, D.: Clustering by passing messages between data points. Sci. **315**(5814), 972–976 (2007)
13. Hankey, W.L.: Re-entry Aerodynamics. AIAA (1988)
14. Hartleib, G.: Tpsx materials properties database. NASA, http://tpsx. arc. nasa. gov
15. Huang, C.H., Galuski, J., Bloebaum, C.L.: Multi-objective pareto concurrent subspace optimization for multidisciplinary design. AIAA J. **45**(8), 1894–1906 (2007)
16. Koziel, S., Ogurtsov, S.: Robust multi-fidelity simulation-driven design optimization of microwave structures. In: Microwave Symposium Digest (MTT), 2010 IEEE MTT-S International, pp. 201–204. IEEE (2010)
17. Lantoine, G., Russell, R.P.: A hybrid differential dynamic programming algorithm for robust low-thrust optimization. In: AAS/AIAA Astrodynamics Specialist Conference and Exhibit (2008)

18. Limbourg, P.: Multi-objective optimization of problems with epistemic uncertainty. In: Evolutionary Multi-Criterion Optimization, pp. 413–427. Springer, New York (2005)
19. Mitcheltree, R., DiFulvio, M., Horvath, T., Braun, R.: Aerothermal heating predictions for mars microprobe. AIAA Paper (98-0170) (1998)
20. Mueller, J.B., Larsson, R.: Collision avoidance maneuver planning with robust optimization. In: International ESA Conference on Guidance, Navigation and Control Systems, Tralee, County Kerry, Ireland (2008)
21. Nguyen, N.V., Choi, S.M., Kim, W.S., Lee, J.W., Kim, S., Neufeld, D., Byun, Y.H.: Multidisciplinary unmanned combat air vehicle system design using multi-fidelity model. Aerospace Science and Technology (2012)
22. Oberkampf, W., Helton, J.C.: Investigation of evidence theory for engineering applications. In: Non-Deterministic Approaches Forum, 43rd AIAA/ASME/ASCE/AHS/ASC Structures, Structural Dynamics, and Materials Conference, pp. 2002–1569 (2002)
23. Ravanbakhsh, A., Mortazavi, M., Roshanian, J.: Multidisciplinary design optimization approach to conceptual design of a leo earth observation microsatellite. In: proceeding of AIAA SpaceOps2008 Conference (2008)
24. Roncoli, R.B., Ludwinski, J.M.: Mission design overview for the mars exploration rover mission. In: 2002 Astrodynamics Specialist Conference (2002)
25. Roshanian, J., Jodei, J., Mirshams, M., Ebrahimi, R., Mirzaee, M.: Multi-level of fidelity multidisciplinary design optimization of small, solid-propellant launch vehicles. Trans. Jpn. Soc. Aeronaut. Space Sci. **53**(180), 73–83 (2010)
26. Saleh, J.H., Mark, G., Jordan, N.C.: Flexibility: a multi-disciplinary literature review and a research agenda for designing flexible engineering systems. J. Eng. Des. **20**(3), 307–323 (2009)
27. Vasile, M.: Robust mission design through evidence theory and multiagent collaborative search. Ann. New York Acad. Sci. **1065**(1), 152–173 (2005)
28. Vasile, M., Minisci, E., Wijnands, Q.: Approximated computation of belief functions for robust design optimization. In: 14th AAA on-Deterministic Approaches Conference (2012) arXiv:1207.3442
29. Vinh, N.X., Busemann, A., Culp, R.D.: Hypersonic and Planetary Entry Flight Mechanics. NASA STI/Recon Technical Report A **81**, 16,245 (1980)
30. Zhang, Q., Zhou, A., Jin, Y.: Rm-meda: a regularity model-based multiobjective estimation of distribution algorithm. IEEE Trans. Evol. Comput. **12**(1), 41–63 (2008)

A Simulation-Based Algorithm for the Probabilistic Traveling Salesman Problem

Weiqi Li

Abstract The probabilistic traveling salesman problem (PTSP) is a variation of the classic TSP and one of the most significant stochastic network and routing problems. Designing effective and efficient algorithms for solving PTSP is a really challenging task, since in PTSP, the computational complexity associated with the combinatorial explosion of potential solution is exacerbated by the stochastic element in the data. In general, researchers use two types of techniques in their search algorithms for PTSP: analytical computation and empirical estimation. The analytical computation approach computes the cost $f(\pi)$ of an a priori tour π using a closed-form expression. Empirical estimation simply estimates the cost through Monte Carlo simulation. This paper describes a simulation-based algorithm that constructs the solution attractor of local search for the PTSP and then finds the best a priori tour within the solution attractor. More specifically, our algorithm first uses a simple multi-start local search process to find a set of locally optimal a priori tours through Monte Carlo simulation and stores these tours in a matrix. Then, the algorithm uses an exhausted search process to find all tours contained in the solution attractor and identifies a globally optimal a priori tour π through Monte Carlo simulation.

1 Introduction

Problem solving under uncertainty has a high impact on the real-world applications, since in the real world many optimization problems are inherently dynamic and stochastic. Such problems exist in many areas such as optimal control, logistic management, dynamic simulation, telecommunications networks, genetics research, neuroscience, and ubiquitous computing. As real-time data in information systems become increasingly available with affordable cost, people have to deal with more and more such complex application problems in which information defining the

W. Li (✉)
University of Michigan, 303 E. Kearsley Street, Flint, MI 48502, USA
e-mail: weli@umflint.edu

M. Emmerich et al. (eds.), *EVOLVE – A Bridge Between Probability,*
Set Oriented Numerics and Evolutionary Computation VII,
Studies in Computational Intelligence 662, DOI 10.1007/978-3-319-49325-1_8

state of the problem continuously changes. Today, uncertainty and dynamism have become much more relevant in many practical applications.

Stochastic combinatorial optimization problems (SCOP) are the optimization problems that include uncertainty in the formulation of the problems. There are two major aspects to consider when modeling optimization problems under uncertainty: first, the way uncertain information is formalized, and second, the dynamicity of the model, that is, the time uncertain information, is revealed with respect to the time at which decisions must be taken [1].

Uncertain information can be formalized in several ways. One way is to describe uncertain information by means of random variables of known probability distributions. Under this model, the optimization problem is stochastic, and the objective function strongly depends on the probabilistic structure of the problem. In SCOPs, we can distinguish a time *before* the actual realization of the random variables and time *after* the random variables are revealed when the associated random events happen. Static SCOPs are characterized by the fact that the identification of a possibly optimal solution is done before the actual realization of the random variables. This model is applicable when a given solution may be applied with no modifications once the actual realization of the random variables is known. This type of optimization problems is known as "a priori" optimization [2, 3]. Dynamic SCOPs arise when it is not possible or not convenient to design a solution that is usable as it is for any realization of the random variables. In this model, optimization efforts must be taken both before and after the random events have happened. In other words, in a dynamic SCOP, a priori solution must be modified dynamically when new information is available [1].

The traveling salesman problem (TSP) is a prototypical combinatorial optimization problem and has many applications in telecommunications, logistics, scheduling, genetics, neuroscience, and other areas. TSP has provided much motivation for development of complexity theory, design of new algorithms, and analysis of solution space [4, 5]. Under the static SCOP framework, the TSP is called *probabilistic TSP*, which consists in finding the a priori tour that visits all cities with minimum expected traveling cost, given that each city has a known probability of requiring a visit. Once the information of which cities actually require a visit in a certain day is known, the cities requiring a visit are visited in the order of the a priori tour, simply skipping the cities not requiring a visit. However, under the dynamic SCOP framework, an a priori tour must be modified dynamically when the uncertain information is reveled. This chapter focuses on the probabilistic TSP.

This chapter introduces a new sampling approximation algorithm to solve the probabilistic TSP, based on the concept of solution attractor of multi-start local search. This chapter is an extended version of [6]. The remaining content of the chapter is organized as follows. Section 2 describes the PTSP. Section 3 discusses the Monte Carlo sampling approximation. Section 4 proposes a simulation-based sampling approximation algorithm, discusses some experimental results on the behavior of the proposed search system, and compares the proposed search system with other PTSP search algorithms. Section 5 concludes this chapter.

2 Probabilistic Traveling Salesman Problem

The classic TSP is defined as: Given a set of n cities and an $n \times n$ cost matrix C in which $c(i, j)$ denotes the traveling cost between cities i and j ($i, j = 1, 2, 3, \ldots,$ $n; i \neq j$). A tour π is a closed route that visits every city exactly once and returns at the end to the starting city. The goal of a search algorithm for the TSP is to find a tour $\pi*$ with minimal traveling cost.

For the TSP under dynamic and stochastic environment, the number of cities n can increase or decrease and the cost $c(i, j)$ between two cities i and j can change with time. This chapter considers only the case in which the number of cities n changes with time t. Therefore, the TSP in the dynamic and stochastic context can be defined as

$$\min \quad f(\pi) = \sum_{i=1}^{n_t - 1} c(i, i + 1) + c(n_t, \ 1) \tag{1}$$

$$\text{subject to} \quad n_t \in N$$

where n_t is the number of cities at time t and N is the set of all potential cities existing in the problem. If we want to design an algorithm to solve the TSP, and the purpose of the algorithm is to continuously track and adapt the changing n through time and to find the currently best solution quickly, that is, to re-optimize the found tour for every change of n, this type of TSP is defined as a dynamic TSP (DTSP). However, if we treat the number of cities n as a random variable and wish to find an a priori tour through all N cities, which is of minimum traveling cost in the expected value sense, this kind of TSP becomes a probabilistic TSP (PTSP). In a PTSP, in any given realization of the problem, the n cities present will be visited in the same order as they appear in the a priori tour, i.e., we simply skip those cities not requiring a visit. The goal of an algorithm for a PTSP is to find a feasible a priori tour with minimal expected cost [3, 7].

The underlying idea of an a priori optimization consists of determining a solution of the whole instance (i.e., the one where all data are present), called an a priori *solution*, and applying a strategy, called a *modification strategy*, that adapts the a priori solution as quickly as possible to the subinstance that must effectively be solved. The choice of this strategy depends strongly on the application modeled by the problem. Typically for a PTSP instance, the first step in a search algorithm is to compute a feasible tour π including the whole set of the cities N; this is an a priori tour π with minimal expected cost. In order to compute the effective tour for a given time, a commonly used *modification strategy* is to drop absent cities and to visit the present cities following the order induced by the a priori tour π [2, 7].

The PTSP was initially introduced by Jaillet [7, 8], who demonstrated that an optimal solution to the deterministic TSP may not be the best solution for a PTSP. Jaillet introduced an analytical framework for the PTSP, examined some of its combinatorial properties and derived a number of asymptotic results. Further, theoretical properties, asymptotic analysis, and heuristic schemes were investigated by some other researchers [2, 9–11]. Surveys of approximation schemes, asymptotic analysis, and complexity theorems for a class of a priori combinatorial optimization problems can be found in [4, 12].

Formally, a PTSP is defined as a complete graph $G = (V, A, C, P)$, where $V = \{v_i: i = 1, 2, \ldots, N\}$ is a set of nodes; $A = \{a(i, j): i, j \in V, i \neq j\}$ is the set of edges that completely connects the nodes; $C = \{c(i, j): i, j \in A\}$ is a set of costs associated with the edges; $P = \{p_i: i \in V\}$ is a set of probabilities that for each node v_i specifies its probability p_i of requiring a visit. In this chapter, the costs in C are assumed to be symmetric; that is, traveling from a node v_i to v_j has the same cost as traveling from node v_j to v_i. The node v_1 is assigned as the depot node with the presence probability of 1. Each non-depot node v_i is associated with a presence probability p_i that represents the possibility that node v_i will be present in a given realization. Based on the values of presence probability (p_i) of non-depot nodes, two types of PTSP can be classified: the homogeneous and heterogeneous PTSP. In the homogeneous PTSP, the presence probabilities of non-depot nodes are all equal ($p_i = p$ for every non-depot node v_i); in the heterogeneous PTSP, these probabilities are not necessarily the same.

Designing effective and efficient algorithms for solving PTSP is a really challenging task, since in the PTSP, the computational complexity associated with the combinatorial explosion of potential solutions is exacerbated by the stochastic element in the problem data [13]. It is clearly much harder to solve a PTSP in practice than a deterministic TSP of the same size. The predominant approaches to finding good solutions for PTSP instances in the literature have been the adaptation of the heuristics for the TSP [9–12]. In general, researchers use two techniques in calculating the cost $f(\pi)$ of an a priori tour π in the PTSP: analytical computation and empirical estimation [8, 13].

Solving the PTSP mainly relies on computing the expected cost of an a priori tour. Theoretically, for a given PTSP instance, the expected cost of an a priori tour can always be computed. A naïve approach calculates the sum of the *a posterior* costs over all possible combinations of realizations of the random variable each multiplied with the according probability that such realizations occur. The computation of the expected cost of a specific a priori tour π for the PTSP instance, denoted $E(\pi)$, depends on the relative location of the cities in that tour and the presence probability of each city in a given instance. By explicitly considering all realizations based on the presence of each individual city, the expected cost of tour π can be calculated. For an N-city PTSP instance, either a city requires a visit or the city does not require a visit, which leads in total to 2^N possible scenarios for N cities, that is, an a priori tour π requires 2^N possible realizations. The probability $p(r_i)$ of realization r_i can be calculated based on the presence probability of each individual city. Let $c[r_i(\pi)]$ describe the cost of tour π for realization r_i under the assumption that cities not in r_i are simply skipped in the tour. The expected tour cost can then be formally described as

$$E(\pi) = \sum_{i=1}^{2^N} p(r_i)c[r_i(\pi)] \tag{2}$$

In general, it is extremely difficult to find an efficient way to compute Eq. 2. Since we have to compute a sum of $O(2^N)$ terms in Eq. 2, it is apparent that the evaluation of $E(\pi)$ is computationally intractable for reasonable size of N, and this naïve computation is therefore not useful for any practical implementation.

An improved method to calculate the exact solution cost analytically is to sum over all edges and multiply their costs with the probability that they occur in the *a posteriori* tour. The probability that a certain edge occurs in the *a posteriori* tour is the product of the probabilities that both of its cities require a visit and that all the cities that are between them in the a priori tour do not require a visit. Thus, the cost $E(\pi)$ can be calculated using the following closed-form expression [7, 8, 14]:

$$E(\pi) = \sum_{i=1}^{N} \sum_{j=i+1}^{N} c_{\pi(i)\pi(j)} p_{\pi(i)} p_{\pi(j)} \prod_{k=i+1}^{j-1} (1 - p_{\pi(k)}) +$$

$$\sum_{j=1}^{N} \sum_{i=1}^{j-1} c_{\pi(j)\pi(i)} p_{\pi(i)} p_{\pi(j)} \prod_{k=j+1}^{N} (1 - p_{\pi(k)}) \prod_{k=1}^{i-1} (1 - p_{\pi(k)}) \tag{3}$$

where $\pi = (\pi(1), \pi(2), \ldots, \pi(N), \pi(N+1) = \pi(1))$ is a permutation of the set V; $c_{\pi(i)\pi(j)}$ represents the cost between cities $\pi(i)$ and $\pi(j)$; $\pi(i)$ denotes the city that has been assigned the ith position in tour π and $p_{\pi(i)}$ is the probability of city $\pi(i)$. Using the Eq. 3, the expected cost can be calculated in run time $O(n^2)$. Although this is much better than the exponential run time of the first approach, it is still too slow for input instances of reasonable sizes.

Unfortunately, designing an effective local search for a PTSP with a computationally expensive objective function is still a challenging task. The reason is that in local search it is crucial to be able to evaluate the neighborhood of a solution efficiently. When the objective function is complex like in Eq. 3, it is difficult to find a delta expression which is both exact and fast to be computed.

Birattari et al. [13] discussed some limitations on analytical computation technique and suggests that the empirical estimation technique can overcome the difficulties posed by analytical computation. Empirical estimation simply estimates the cost $E(\pi)$ of an a priori tour through Monte Carlo simulation. Instead of summing over all possible scenarios, we could sample M ($M < 2^N$) scenarios of using the known probabilities and take the average over the costs of the *a posteriori* tours for the sampled scenarios.

In recent years, several search algorithms, using analytical computation and/or empirical estimation approach, have been proposed to solve the PTSP. Laporte et al. [15] provided an exact algorithm for the PTSP. They used an integer L-shaped method and solved to optimality instances involving up to 50 cities. However, the exact approach is limited to small problem sizes. Consequently, much of the PTSP literature focuses on heuristic and meta-heuristic approaches.

Bertsimas et al. [3] investigated some properties of the PTSP and proposed some tour construction heuristics and tour improvement heuristics for the PTSP. Bertsimas and Howell [9] introduced the 2-p-opt local search and the 1-shift local search for the PTSP, using new equations for efficiently evaluating the cost of local search moves. Later, Bianchi et al. [16] provided corrections for the equations in [9]. Bianchi and Campbell [17] extended the 2-p-opt and 1-shift local search to the heterogeneous PTSP. Liu [18] introduced a concept of diversified local search strategy under the scatter search framework for the PTSP. Weyland et al. [19] presented 3-opt sampling local search algorithm using delta evaluation and the Monte Carlo sampling-based approximation to evaluate solutions. Birattari et al. [13] introduced an estimation-based iterative improvement algorithm, called 2.5-opt-EEs, that performs delta evaluation using empirical estimation techniques. Marinakis et al. [20] proposed an expending neighborhood search GRASP method for the PTSP. Campbell [21] presented an aggregation method for the solution of the PTSP. Marinakis et al. [22] provided a stochastic dynamic programming algorithm for the PTSP.

To efficiently and effectively solve the PTSP, recent studies have focused on adopting new algorithmic approaches based on meta-heuristics such as simulated annealing (SA), evolutionary algorithm (EA), genetic algorithm (GA), and ant colony optimization (ACO). Bowler et al. [23] used a stochastic SA algorithm to experimentally analyze the asymptotic behavior of suboptimal homogeneous PTSP solution, in which the objective function is estimated by sampling and the sampling estimation error is used instead of the annealing temperature. Liu [24] considered a hybrid scatter search EA for the PTSP that incorporates the use of the nearest neighbor constructive heuristic, threshold accepting screening mechanism, and crossover operator. Liu [25] also proposed an optimization procedure based on GA that incorporates the nearest neighbor algorithm, 1-shift and/or 2-opt exchanges for local search, selection scheme, and edge recombination crossover operator into genetic algorithm framework. Liu et al. [26] used an EA with diversified crossover operator to solve the heterogeneous PTSP. Bianchi et al. [27, 28] investigated the potentialities of ACO algorithms for both homogeneous and heterogeneous PTSP under different probability configurations of cities. Gutjahr [29, 30] proposed and analyzed a sampling-based ACO algorithm, in which statistical tests are performed for comparing the sample average values of the solutions generated by ants in order to select a set of best solutions. Branke and Guntsch [31] proposed two modifications to the standard ACO meta-heuristic which enhance its performance for the PTSP. They first examined the use of an approximation of the evaluation function for solutions in order to save time that could then be used for constructing more solutions; second, they devised new heuristic guidance schemes for the ants which more accurately reflect the impact of a decision on solution quality. Birattari et al. [32, 33] proposed an empirical estimation-based ACO/F-Race algorithm, where at each iteration the selection of the new best solution is done with a procedure called F-Race. F-Race consists in a series of steps at each of which a new scenario is sampled and is used for evaluating the solutions that are still in the race. At each step, a Friedman test is performed and solutions that are statistically dominated by at least another one are discarded from the race. Liu [34] developed a memetic algorithm (MA) by incorporating the nearest neighbor algorithm to generate initial solutions, 1-shift and/or 2-opt exchanges for local search, and edge recombination crossover to efficiently and effectively solve the PTSP. Marinakis and Marinakis [35] designed a hybrid multi-swarm optimization algorithm for the PTSP. A comprehensive review on the developments in the meta-heuristic algorithms filed can be found in [1].

3 Monte Carlo Sampling Approximation

Stochastic optimization problems involve integrals of any probabilistic function. Owing to the intractability of exactly solving these complex problems, in recent years, a great deal of attention has been devoted to theoretical and practical aspects of combining optimization and simulation techniques to solve these types of optimization problems [36].

The sample average approximation (SAA) method is an approach for solving stochastic optimization problem by using simulation. In this method, the expected objective function of the stochastic problem is approximated by a sample average estimate derived from a sample. The given stochastic optimization problem is therefore transformed into a so-called *sample average optimization problem*. The resulting sample average approximating problem is then solved by deterministic optimization techniques. The expected value of the objective function is obtained by considering several realizations of the random variable and by approximating the cost of a solution with a sample average function [37]. The process is repeated with different samples to obtain candidate solutions along with statistical estimates of their optimality gaps. Hence, a sampling technique which provides a representative sample from the multivariate probability distribution is crucial in obtaining true performance statistics for optimization. Here, it is assumed that the uncertainty distributions are a priori known. The SAA method has been used successfully to solve stochastic optimizations with a large sample of realizations [1, 37–40].

In the SAA method, the objective value of a new solution is computed by a sample average of the type of the following equation:

$$G_M(x) := \frac{1}{M} \sum_{i=1}^{M} G(x, \omega_i) \tag{4}$$

The objective function $G_M(x)$ is typically approximated by the sample average. $\omega_1, \omega_2, \ldots, \omega_M$ is a random sample of M independent, identically distributed realizations of the random vector ω. The sample average is also referred to as sample estimate, and the random realizations referred to as random scenarios.

In sampling approximation search, a stochastic optimization problem is usually represented by a computer simulation model. Simulation models are models of real or hypothetical systems, reflecting all important characteristics of the system under studied. Optimization via simulation means searching for a solution that yields the maximum or minimum expected value of the problem that is represented by a simulation model. One of the best known methods for sampling a probability distribution is the Monte Carlo sampling technique, which is based on the use of a pseudo-random number generator to approximate a uniform distribution.

Currently, the Monte Carlo sampling approximation method is the most popular approach for solving stochastic optimization problems. This method approximates the expected objective function of the stochastic problem by a sample average estimate derived from a random sample. We assume that the sample used at any given iteration is independent and identically distributed and that this sample is independent of previous samples. The resulting sample average approximating problem is then solved by deterministic optimization techniques. The process can be repeated with different samples to obtain candidate solutions along with statistical estimates of their optimality gaps. Sampling approximation via simulation is statistically valid in the context of simulation as the underlying assumptions of normality and independence of observations can be easily achieved through appropriate sample averages of independent realizations, and through adequate assignment of the pseudo-random number generator seeds, respectively [37].

In the case of PTSP, the elements of the general definition of the stochastic problem take the following format: A feasible tour π is an a priori tour visiting once and only once all N cities, and the random variable n is extracted from an N-variate Bernoulli distribution and prescribes which cities need being visited. The cost $f(\pi)$ of a PTSP tour π can be empirically estimated on the basis of a sample costs $f(\pi, n_1), f(\pi, n_2), \ldots, f(\pi, n_M)$ of a posteriori tours obtained from M independent realizations $n_1, n_2, \ldots, n_M$ of the random variable n:

$$\hat{f}_M(\pi) = \frac{1}{M} \sum_{i=1}^{M} f(\pi, n_i) \tag{5}$$

$\hat{f}_M(\pi)$ denotes the average of the objective values of the M realizations on the a priori tour π, which gives us an approximation for the estimated cost for tour π. Clearly, $\hat{f}_M(\pi)$ is an unbiased estimator of $f(\pi)$. A search algorithm for PTSP is looking for the optimal tour π^* which has the smallest estimated objective value $\hat{f}_M(\pi^*)$, that is,

$$\pi^* \in \arg \min\{\hat{f}_M(\pi_1), \hat{f}_M(\pi_2), \ldots\} \tag{6}$$

The optimal value $\hat{f}_M(\pi^*)$ and the optimal tour π^* to the PTSP provide estimates of their true counterparts.

One of the main disadvantages of the Monte Carlo method is that the bound is probabilistic, and there is no methodical way for constructing the sample points to achieve the probabilistic bound. Therefore, optimization via simulation adds an additional complication because the value of a solution cannot be evaluated exactly, but instead must be estimated. Because we have estimates, it may not be possible to conclusively determine if π_i is better than π_j, which may frustrates the search algorithm that tries to move in an improving direction. In principle, we can eliminate this complication by making so many replications at each iterative point that the performance estimate has essentially no variance. In practice, this could mean that we will explore very few iteration due to the time required to simulate each one. Therefore, in a practical sampling approximation algorithm, the test if a solution is better than another one can only be done by statistical sampling, that is, obtaining a correct comparison result only with a certain probability. The goal now is to get a good average case solution and the expected value of the objective is to be optimized. The way simulation approximation is used in an optimization algorithm largely depends on the way solutions are compared and the best solutions among a set of other solutions is selected.

A major determinant of the computational cost for a simulation-based optimization algorithm is the number of simulation replications used to estimate the cost for each a priori tour in PTSP. A key feature that is not a factor in classic TSP is the trade-off between the amount of computational effort needed to estimate the cost for a particular solution versus the effort in finding improved solutions. The number of realizations M should be large enough for providing a reliable estimate of the costs of solutions, but at the same time it should not be too large otherwise too much time is wasted. The appropriate number of realizations M depends on the stochastic character of the problem at hand. The larger the probability that a city is to be visited, the less stochastic an instance is. In this kind of case, an algorithm can be designed to consider a reduced number of realizations and therefore explore more solutions in the unit of time. On the other hand, when the probability that a city is to be visited is small, the instance at hand is highly stochastic. In such a case, it pays off to reduce the total number of solutions explored and to consider a large number of realizations for obtaining more accurate estimate [37, 41, 42].

There are several sampling strategies available. For the PTSP, popular sampling strategies include (1) the same set of M realizations is used for all steps of the iteration in the algorithm; (2) a set of M realizations is sampled anew each time an improved solution is found; and (3) a set of M realizations is sampled anew for each comparison of solutions. The first strategy is a well-known variance-reduction technique called the method of common random numbers (CRN). CRN takes advantage of the same set of random numbers across all alternatives for a given replication. CRN is typically designed to induce positive correlation among the outputs of each respective alternative for a given replication, thereby reducing the variance of the difference between the mean alternative point estimators. One of the practical motivations for using CRN in a search algorithm is to speed up the sample average computations. However, one major problem with CRN is that the iterations of the algorithm may be "trapped" in a single "bad" sample path. Second and third strategies are called variable-sample method. In variable-sample method, the objective function is replaced, at each iteration, by a sample average approximation. This resampling technique allows the iterations of the algorithm to get away from the possible "bad" sample paths. Another advantage of a variable-sample scheme is that the sample sizes can increase along the algorithm, so that sampling effort is not wasted at the initial iteration of the algorithm [36].

Some researchers have been proposed estimation-based algorithms to deal with the PTSP, using local search or meta-heuristics [19, 36, 40–43]. This chapter introduces a new optimization approach that constructs solution attractor of multi-start local search in the context of Monte Carlo simulation. This new algorithm combines optimization and simulation in a parallel iterative process in order to gain the advantages of optimization (exact solution), simulation (stochasticity), and speed (parallel processing).

4 The Proposed Simulation-Based Search Algorithm

4.1 The Parallel Search System

The proposed search system uses a parallel multi-start local search procedure to construct the solution attractor for the PTSP. Then, the optimal solution is found in the solution attractor. In the search system, the costs of solutions are estimated using Monte Carlo simulation. This section describes the parallel attractor-construction procedure for the PTSP.

A search trajectory is the path followed by the search process in the solution space as it evolves with time. A multi-start heuristic search algorithm produces several search trajectories and generates different locally optimal solutions. The solution attractor of a multi-start search algorithm is defined as a subset of the solution space that contains the whole solution space of the end points (locally optimal solutions) of all local search trajectories. The solution attractor drives the search trajectories to converge into a small region in the solution space [44]. Figure 1 illustrates the concepts of local search trajectories and solution attractor of a multi-start local search system. Figure 2 presents a parallel algorithms for constructing the solution attractor of local search for an n-city TSP instance implemented in a master–worker architecture. This attractor-construction procedure is very straightforward: generating M locally optimal tours, storing them into an $n \times n$ matrix (called *hit-frequency matrix* $E = \{e_{ij}\}_{n \times n}$ and then finding all tours contained in the matrix E. In the procedure, C is the cost matrix for a TSP instance G, π_i is an initial tour generated by the

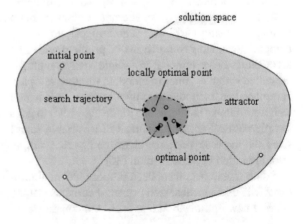

Fig. 1 The concepts of search trajectories and solution attractor in a multi-start local search system

```
procedure TSP_Attractor_Master(G)
begin
  initialize E;
  for i = 1 to K
    send C to worker_processor_i;
  end for
  repeat
  receive a tour from a worker_processor;
  update E;
  until number_of_local_optima = M
  send Stop_signal to K worker_processors;
  Exhausted_Search(E);
end

procedure TSP_Attractor_Worker_i(C)
begin
  repeat
    π_i = Initial_Tour();
    π_j = Local_Search(π_i);
  send π_j to master_processor;
  until Stop_signal
end
```

Fig. 2 The parallel procedure for constructing solution attractor of multi-start local search in TSP with master–worker implementation

function Initial_Tour(). The function Local_Search() runs a local search on π_i and outputs a locally optimal tour π_j. The hit-frequency matrix E is used to record the number of hits on the edge $a(i, j)$ of the graph by the set of M locally optimal tours. In principle, the matrix E can be the architecture that allows individual locally optimal tours to be linked together along a common structure and generate the structure of the solution attractor. This solution attractor can help to find the globally optimal tour [44]. Li [45–47] applied the solution attractor concept to tackle multi-objective TSP and dynamic TSP. This chapter applies this concept to solve PTSP.

Because the procedure for constructing solution attractor of multi-start local search can be divided in such a way that separate CPUs can start and execute different search trajectories without interfering with each other, we implemented our proposed multi-start search algorithm into a parallel search system. Parallel multi-start search explores different areas in the solution space at the same time; thus, it generates a wide sample of the local optima. This type of parallel processing can be useful to efficiently solve difficult optimization problems, not only speeding up the execution times, but also improving the quality of the final solutions.

Figure 3 sketches the basic framework of our parallel multi-start search system for PTSP. This search system contains $K + 1$ computers and bears intrinsic parallelism in its data and processing structures. Based on a common cost matrix C and a probability array P, this search system starts K separate search trajectories in parallel. When a search trajectory reaches its locally optimal point, the processor stores the solution in the common hit-frequency matrix E. Then, the processor starts a new search if more computing time is available. Finally, at the end of the search, the matrix E is searched

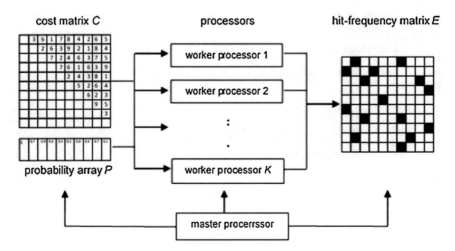

Fig. 3 Schematic structure of the parallel search system for PTSP

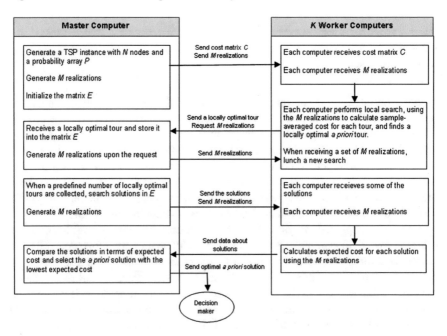

Fig. 4 The parallel search system implemented in a master–worker architecture

by an exhausted search process and the best solution in the attractor is outputted as the optimal solution.

Figure 4 illustrates the implementation of the proposed parallel search system using a master–worker architecture. Figures 5 and 6 lists the procedures for master computer and worker computers. In this master–worker parallel search system, one

```
procedure PTSP_Master()
begin
  generate a problem instance with N nodes;
  generate a probability array P;
  generate M realizations on N according to P;
  initialize the matrix E;
  for i = 1 to k
    send cost matrix C and M realizations to worker i;
  end for
  repeat
    receive a locally optimal tour from a worker;
    store the tour into E;
    generate M realizations upon request and sent to the worker;
  until predefined_number_of_tours
  send stop_signal to workers;
  L = Exhausted_Search(E);
  divide the tours in L into k groups;
  generate M realizations;
  for i = 1 to k
    send L_i and M realizations to worker i;
  end for
  repeat
    receive a tour with expected cost value from a worker;
  until all_tours_received
  compare the tours in terms of expected costs;
  output the best a priori tour;
end
```

Fig. 5 The algorithm for master computer

computer serves as a master and other computers as workers. The master computer generates a TSP instance with N nodes and a probability array $P = \{p_i\}_N$, initializes the hit-frequency matrix E, and sends a copy of the cost matrix C to each of the worker computers. The multi-search task is distributed to K worker computers.

Then, based on the probability values in the array P, the master computer generates a set of M realizations $(m_1, m_2, \ldots, m_M)$. Each of the M realizations is an array that contains binary values, where a value "1" at position i in the array indicates that node v_i requires being visited whereas a value "0" means that it does not require being visited. The master computer then sends the set of M realizations to the worker computers. Depending on the implementation setting, the master computer can send the same set of M realizations to all worker computers or a different set of M realizations to each of the worker computers. In our implementation, the master computer sends the same set of M realizations to all worker computers.

```
procedure PTSP_Workerᵢ()
  begin
  receive cost matrix C from the master;
  while (stop_signal not true)
    receive M realizations from the master;
    s = Local_Search(C, M);
    send s to the master;
    request a new set of M realizations from the master;
  end while
  receive Lᵢ tours from the master;
  receive M realizations from the master;
  for i = 1 to Lᵢ
    calculate expected cost uᵢ for tour sᵢ using M realizations;
    send sᵢ with uᵢ to the master;
  end for
end
```

Fig. 6 The algorithm for worker computers

When receiving a set of M realizations, each worker computer independently performs its local search: it randomly generates an initial a priori tour, calculates the sample-averaged cost by using the M realizations; it then generates a new a priori tour and calculates its sample-averaged cost by using the same M realizations; if this new a priori tour has lower average cost, the current a priori tour is replaced by this better a priori tour; otherwise, another new a *priori* tour is generated and compared with the current a priori tour; when the search process reaches a locally optimal a priori tour, the worker computer sends the locally optimal tour to the master computer. Figure 7 presents the algorithm of the local search process with simulation. After sending a locally optimal tour to the master computer, the worker computer can start a new search. The worker computer can use the same set of M realizations or a new set of M realizations in the new search. In our implementation, when a worker computer starts a new search, it requests the master computer to send a new set of M realizations and then evaluates the tours using the new M realizations. In other words, each locally optimal tour is generated by using a different set of M realizations.

When the master computer receives a locally optimal a priori tour from a worker computer, it stores the tour into the matrix E. It then generates a new set of M realizations and sends it to the requesting worker computer. When the worker computer receives this new set of M realizations, it lunches a new local search. In such a way, the multi-start search is performed by the K worker computers in parallel.

When a predefined number of locally optimal a priori tours are collected from the worker computers and stored in the matrix E, the master computer sends a stopping signal to the worker computers to stop their search processes. The set of locally optimal tours in E forms a solution attractor for the problem.

```
procedure Local_Search(C, M)
begin
  s₁ = Generate();
  total_cost = 0;
  for i = 1 to M
    calculate cost cᵢ for mᵢ realization;
    total_cost = total_cost + cᵢ;
  end for
  u₁ = total_cost / M;
  repeat
    s₂ = Generate();
    total_cost = 0;
    for i = 1 to M
      calculate cost cᵢ for mᵢ realization;
      total_cost = total_cost + cᵢ;
    end for
    u₂ = total_cost / M;
    if (u₂ < u₁)
      u₁ = u₂;
      s₁ = s₂;
    end if
  until stop_criteria
  return s₁, u₁;
end
```

Fig. 7 The algorithm of local search using simulation

Then, the master computer launches an exhausted search process in the matrix E and identifies all tours in E. These tours are then divided into K groups $L_1, L_2, \ldots, L_k$. The master computer generates a set of M realizations and send a group of tours L_i with the M realizations to i worker computer. The worker computer i uses the set of M realizations to calculate the sampling-average cost for each of the tours in L_i and send the cost value with the tour back to the master computer. The master computer compares all tours in terms of sampling-average costs and outputs the a priori tour with lowest average cost.

In the proposed search algorithm, the Monte Carlo simulation is used in two levels: at the local search level, the worker computers use the Monte Carlo simulation to calculate the sampling-average cost for each tour during the search; and at the global search level, the worker computers use the Monte Carlo simulation to calculate the sampling-average cost for each of tours in the solution attractor and the master computer chooses the globally optimal tour that has the smallest estimated cost.

4.2 The Experimental Setting

PTSP is a new research and application area in the TSP class. Due to its novelty, so far there is no test problem instance available to be used for implementing a new search algorithm and assessing its suitability, quality, effectiveness, and efficiency. Therefore, we design a test problem for the implementation and evaluation of our proposed algorithm.

The design of a test problem is always important in designing any new search algorithm. The context of problem difficulty naturally depends on the nature of problems that the underlying algorithm is trying to solve. In the context of solving PTSP, we designed our test problem based on several considerations. First, the size of problem should be large because the classic TSP instances as small as 200 cities are now considered to be well within the state of the global optimization art. A problem instance must be considerable large than this for us to be sure that the heuristic approach is really called for. Second, there is no any pre-known information related to the results of the experiment, since in a real-world problem one does not usually have any knowledge of the solutions. There is no motivation for an algorithm designer to fine-tune the proposed search algorithm in order to make the results of the algorithm look better. Third, when dealing with stochastic combinatorial problems, randomness in both process and data means that the underlying model in the algorithm must be suitable for the modeling of natural stochastic problems. In other words, we should formulate a problem that goes nearer to real-world stochastic conditions. Last, the problem instance should be general, understandable, and easy to formulate so that the experiments are repeatable and verifiable.

We generated a TSP instance with $N = 1000$ cities. The cost matrix C was generated at random, where each cost element $c(i, j)$ in C was assigned a random integer number independently drawn from an uniform distribution of the integers in the range [1, 1000]. Our TSP instances were a general symmetric TSP, with only $c(i, j) = c(j, i)$ restriction. The triangle inequality $c(i, j) + c(j, k) \geq c(i, k)$ was not assumed in our problem instance. A probability array P contained a set of probability values that assigned probability p_i to city i for requiring a visit. We specified node 1, v_1, as the depot node with $p_1 = 1.0$. The probability for each of non-depot cities was generated from a uniform random number in a certain range. Therefore, our test problem is a heterogeneous PTSP.

Our search algorithm was implemented using Sun Java JDK 1.3.1 and all experiments were conducted on a network of 6 PCs. Each PC was with Pentium 4 at 2.4 GHz and 512 MB of RAM running Linux. The PCs were interconnected with a fast Ethernet communication network using the LAM implementation of the MPI standard. The network was not dedicated, but was very steady. In our parallel search system, one computer served as a master computer and other five computers as worker computers. Our network architecture and algorithms were asynchronous by nature, meaning that the processors did not have a common clock by which to synchronize their processing and calculation.

Our experiments relied heavily on randomization in order to generate a wide sample of locally optimal tours. All initial tours were randomly constructed. We used simple 2-opt local search in our search system. 2-opt is an iterative improvement technique, which starts from some initial solution and then moves to an improving neighboring solution by exchanging two edges until a local optimum is found. During the local search, the search process randomly selected a solution in the neighborhood of the current solution. A move that gave the first improvement was chosen. The great advantage of first-improvement pivoting rule in search process is to produce randomized local optima. The local search process on each search trajectory terminated when no improvement could be achieved after 10000 iterations. We assume that the sample of M realizations used at any given iteration is independent and identically distributed and that this sample is independent of previous samples.

4.3 The Experimental Analysis

A main challenge in applying local search to PTSP lies in designing an effective evaluation procedure that conclusively determines if one tour is better than another. Because our search system uses simulation and sampling approximation, it is not possible to decide with certainty whether a tour is better than another during the search. This type of comparison can only be tested by statistical sampling, obtaining a correct comparison result only with a certain probability. In other words, the simulation estimates should be accompanied with some indication of precision. The first decision we had to make in our experiment was to choose an appropriate sample size M (number of realizations) in our simulation. The number of realizations considered should be large enough for providing a reliable estimate of the cost of tours but at the same time it should not be too large otherwise too much computing time is wasted. In our experiments, the accuracy factors we considered include desired precision of results, confidence level, and degree of variability. We used the following equation to determine our sample size M [48, 49]:

$$M = \frac{V(1-V)}{\frac{A^2}{Z^2} + \frac{V(1-V)}{P}} \qquad (7)$$

where M is the required sample size; V is the estimated variance in population, which determines the degree to which the attributes being measured in the problem are distributed throughout the population; A is the desired precision of results that measures the difference between the value of the sample and the true value of the real population, called the sampling error; Z is the confidence level that measures the percentage of the samples would have the true population value within the range of chosen precision; and P is the size of population. In our case, the size of population is 2^{1000}.

In our search system, we have two levels of simulation. At the local search level, the worker computers use simulation to calculate sample averages and use the averages to compare tours. At this level, we choose the sampling error $A = \pm 3\%$ and confidence level $Z = 1.96$ (95 % confidence). Because our PTSP instance is heterogeneous, we use variability $V = 50\%$. Using Eq. 7, we calculate $M = 1067$. At the global search level, the workers use simulation to calculate sample averages for the tours found in the matrix E, and then, the master computer uses this information to order the tours and select the best one. At this level, we choose the sample error $A = \pm 3\%$, confidence level $Z = 2.57$ (99 % confidence) and $V = 50\%$. We calculate $M = 1835$. Therefore, in our experiment, we used $m_{local} = 1100$ realizations in the local search level and $m_{global} = 1850$ in the global search level.

In one experiment, the master computer generated a TSP instance and a probability array P, in which the value of p_i was generated from a uniform random number in the range $[0.1, 0.9]$. Then, the master computer generated a set of 1100 realizations based on P and sent this set to all worker computers. The worker computers performed local search and used the 1100 realizations to calculate the sampling-average cost for each of tours. When a worker computer found a locally optimal tour, it sent the tour to the master computer. When the master computer collected 30 locally optimal tours from the worker computers, it sent a stopping signal to the worker computers to stop local searching. The master computer then applied an exhausted search procedure on the matrix E and found 36 tours in E. The master computer sent these tours to the worker computers. Four worker computers received 7 tours each and one worker computer received 8 tours. The master computer generated a set of 1850 realizations and sent it to all worker computers. The worker computers used this new set of realizations to calculate the sampling-average costs and standard deviations for these 36 tours. Finally, the master computer used these sampling-average costs to order these tours. Table 1 lists the five best tours found in E. We can see that the expected cost of the best tour is 4889 with standard deviation 201.

Then, we ran the search system on the same PTSP instance four more times. Each time, the search system used the same probability array P, but generated different sets of realizations for m_{local} and m_{global}. Table 2 lists the results of these five trials. The table shows the number of tours found in the matrix E, the sampling-average cost of the best tour found in each of the trials and total computing time consumed by each of the computers. We compared these five tours and found that they were actually the same tour; even it had a different sampling-average value in each of trials. This best tour is probably the globally optimal *a prior* tour for the PTSP instance.

Table 1 The five best tours found in the matrix E

Tour	Sampling average	Standard deviation
1	4889	201
2	4902	224
3	5092	199
4	5413	213
5	5627	209

Table 2 Results of five trials on the same PTSP instance

Trial	Number of tours in E	Sampling average for best tour	Standard deviation	Time (s)					
				Master	Worker 1	Worker 2	Worker 3	Worker 4	Worker 5
1	36	4889	201	1007	1578	1549	1561	1468	1593
2	35	4872	209	1220	1563	1600	1589	1564	1465
3	36	4850	195	1157	1480	1533	1474	1593	1572
4	34	4873	199	1004	1550	1472	1543	1569	1530
5	35	4880	202	1058	1461	1486	1542	1473	1609

A different experiment studied the effect of the problem stochastic nature (i.e., different probability P) on the quality of search in our search system. The results of this experiment are shown in Table 3. The experiment used the same TSP instance but three different probability arrays P_1, P_2, and P_3. The values in P_1, P_2, and P_3 were generated from the range [0.1, 0.4], [0.3, 0.7] and [0.6, 0.9], respectively. $m_{local} = 1100$ and $m_{global} = 1850$ were still used in the experiment. For the probability array P_1, the experiment ran the search system five times, each time using different sets of m_{local} and m_{global}. These five trials generated three different best tours: first and third trials generated the same best tour with different sampling-average values; second trial generated a different best tour; and fourth and fifth trials generated the same best tour that was different from the one generated by first three trials. This fact indicates that, when a PTSP instance becomes more stochastic, our search system has more difficulty to find the globally optimal a priori tour. Then, the experiment ran the search system five times for the probability array P_2, the five trials generated the same best tour with different sampling-average values. Last the experiment ran the search system five times for the probability array P_3, the search system also outputted the same best tour in the five trials. Obviously, when a PTSP instance is less stochastic or its average probability is 50/50, our search system is able to find the globally optimal a priori tour. This experiment indicates that the "level of stochasticity" of a problem instance is an important factor that affects the quality of search systems.

For the PTSP instance with the probability array P_1, next experiment wanted to know if the search quality could be improved by increasing the number of realizations in the simulation process. The experiment first used $m_{local} = 1100$ and $m_{global} = 1850$ and ran the search system on the problem instance ten times. Five different best tours were generated in the ten trials and five trials generated the same best tour. Then, the experiment used $m_{local} = 3000$ and $m_{global} = 3000$, ran the search system on the problem instance ten times again. Three different best tours were generated by the ten trials and seven trials generated the same best tour. Then, the experiment did the same procedure using $m_{local} = 6000$ and $m_{global} = 6000$. This time two different best tours were generated and nine trials gave us the same best tour. This experiment indicates that the search quality can be improved by increasing the number of realizations in the simulation process. Table 4 summaries the experiment results.

4.4   Performance Comparison with Other Algorithms

This subsection presents the performance comparison results between the proposed search system and some other current state-of-the-art approaches. The proposed search system was compared with four search algorithms: 2-p-opt, 1-shift, 2.5-opt-EEs, and pACS. They are all straightforward adaptations of heuristics for the classical TSP and used to tackle the PTSP.

Table 3 The results of experiment on different probability arrays

Trial	Number of tours in E	Sampling average for best tour	Standard deviation	Time (s)					
				Master	Worker 1	Worker 2	Worker 3	Worker 4	Worker 5
P_1									
1	39	1201	275	968	1261	1319	1208	1232	1304
2	37	1180	271	992	1290	1302	1258	1244	1326
3	39	1260	269	985	1239	1211	1215	1309	1269
4	42	1238	267	965	1237	1283	1278	1319	1211
5	39	1189	279	956	1309	1256	1304	1251	1245
P_2									
1	37	5580	166	993	1492	1345	1491	1454	1330
2	37	5606	175	980	1428	1381	1444	1307	1340
3	39	5554	164	1063	1465	1142	1313	1342	1493
4	36	5581	171	1044	1429	1384	1397	1475	1430
5	37	5561	161	995	1432	1369	1481	1330	1382
P_3									
1	34	7047	156	1201	1502	1558	1561	1563	1478
2	35	7049	154	1156	1583	1496	1558	1537	1525
3	33	7012	165	1127	1570	1487	1476	1471	1561
4	36	7009	161	1211	1495	1474	1589	1535	1589
5	37	7108	163	1124	1533	1505	1538	1538	1574

Table 4 Search results in three different realizations M

Setting	Number of different best tours in 10 trials	Number of trials having the same tour	Average standard deviation
$M_{local} = 1100$ $M_{global} = 1850$	5	5	268
$M_{local} = 3000$ $M_{global} = 3000$	3	7	247
$M_{local} = 6000$ $M_{global} = 6000$	2	9	226

The 2-p-opt and 1-shift local search algorithms were introduced by Bertsimas in [2, 9]. The 2-p-opt neighborhood of an a priori tour is the set of tours obtained by reversing a section of the a priori tour. It is the probabilistic version of the famous 2-opt algorithm created for the classical TSP. The 1-shift neighborhood of an a priori tour is the set of tours obtained by moving a node which is at position i to position j of the tour, with the intervening nodes being shifted backward one space accordingly. In our comparison with the 2-p-opt and 1-shift, we use the correct delta objective evaluation expression introduced in [16].

The 2.5-opt-EEs is an estimation-based local search algorithm [13]. A particularity of 2.5-opt-EEs is that the cost of the neighbor solutions is calculated using delta evaluation through empirical estimation. A description of the 2.5-opt neighborhood operator, which is a combination of the 2-opt and 1-shift neighborhood operators, can be found in [50]. A detailed description for 2.5-opt-EEs can be found in [51].

The pACS is a modified algorithm to apply ant colony system to the PTSP [38]. We use the same parameter settings for the pACS described in [38], without tuning them to our particular problem instances. That is, the parameters for the pACS are the number of ants $m = 10$, the probability that the next customer is chosen deterministically $q_0 = 0.95$, the power of heuristic information exponent $\beta = 3$, the local evaporation factor $\rho = 0.1$, and the global evaporation factor $\alpha = 0.1$. The objective function value is estimated using sampling approximation. The number of samples M is kept fixed through all the iterations.

The comparison experiment was conducted on two heterogeneous PTSP instance with $N = 1000$ and $N = 2000$, each with four different probability arrays $P_1 = [0.1, 0.4]$, $P_2 = [0.3, 0.7]$, $P_3 = [0.6, 0.9]$, and $P_4 = [0.1, 0.9]$. We used $m_{local} = m_{global} = 2000$ for our search system on both instances, and sample $M = 2000$ for the 2.5-opt-EEs and pACS algorithms. In order to reduce variance, the same set of M realizations was used for all steps of a particular algorithm.

Each instance was tested by running each of the algorithmic procedures 50 times using different random seeds and averaging the results in an attempt to enhance the robustness of the results. In local search part, the tours were explored in a random order in each iteration and the first improving tour was used to replace the current tour. A locally optimal tour was reached when no improvement was made during 10000 iterations.

Table 5 The comparison results on two PTSP instances for different probability ranges

N	P	$E(\pi)$				
		Proposed system	2-p-opt	1-shift	2.5-opt-EEs	pACS
1000						
	[0.1, 0.4]	1749*	1893	1874	1865	1863
	[0.3, 0.7]	4256*	4426	4463	4378	4315
	[0.6, 0.9]	6947*	7125	7233	7096	7119
	[0.1, 0.9]	5581*	5686	5672	5664	5653
2000						
	[0.1, 0.4]	3215*	3417	3401	3389	3385
	[0.3, 0.7]	6855*	7026	7128	6995	7012
	[0.6, 0.9]	8936*	9190	9334	9181	9213
	[0.1, 0.9]	7427*	8834	8865	7854	7925

Each algorithm was allowed to run until it reached a locally optimal point. The results were obtained in average over 50 trials for each instance in each probability range. Table 5 presents the comparison results. N denotes problem size; P represents the customer presence probability interval; and $E(\pi)$ denotes the average value of the expected cost of the a priori PTSP tour over 50 trials. The best value in each line is bold, with an asterisk "*" indicating whether the difference to the closest competitor is significant according to a t test with a confidence of 95 %. Table 5 does not include CPU running time for comparison test, because our search system was carried out in a parallel multi-processors network and other algorithms were performed in individual computers with different computing platforms.

As shown in Table 5, our proposed search system consistently performs better in both PTSP instances for all probability range in terms of the expected cost $E(\pi)$ in a statistically significant using t test with 95 % confidence. The observed better performance of our search system can be due to its capability of finding the globally optimal solutions.

5 Conclusion

This chapter introduces a new search algorithm for PTSP and describes the parallel implementation of the algorithm. The main novelty of the proposed approach is the use of solution attractor concept for global solutions and use of the empirical estimation technique for evaluating their values. In our algorithm, the solution attractor of local search is used to obtain globally optimal solutions and Monte Carlo simulation is used to evaluate the solutions for the stochastic problem. Through multi-start search and multiple parallel processors, our search system bears intrinsic diversity, flexibility, parallelism, and global nature.

In our search system, all samples of realizations are generated by the master computer in order to reduce the variances of the estimators and computational effort. However, a parallel approach may be advantageous in which realization samples are derived in parallel from different processors. When each simulation is performed in a different processor, the search system can be really ergodic. Taking over a large ensemble of Monte Carlo salesmen, each of them trapped in some local valley, we may get more correct values for the average of the estimator.

The pervasive nature of information technology in modern society has sparked the development of novel computing schemes in which the resources of many separate computers connected by a network are used to solve large-scale computational problems. Our proposed search system can be scalable to large-scale instance sizes. For example, our algorithm can effectively use grid computing model in which a large number of computers form a virtual cluster to tackle large-scale complex optimization problems in a massively parallel fashion.

This chapter compared our proposed search system with other PTSP algorithms. The comparison result demonstrates that our search system outperforms other PTSP algorithms. Of course, a fair comparison with other algorithms on stochastic problem is almost impossible, since the solution quality depends heavily on the level of stochasticity of the problem instance used in the search, and the bound of Monte Carlo method is also probabilistic.

The PTSP is a basic stochastic optimization problem. The results from the study of the PTSP can provide insights into research in other probabilistic combinatorial optimization problems, the potential areas for future research.

References

1. Bianchi, L., Dorigo, M., Gambardella, L.M., Gutjahr, W.J.: A survey on metaheuristics for stochastic combinatorial optimization. Natural Comput. **8**, 239–287 (2009)
2. Bertsimas, D.J.: Probabilistic Combinatorial Optimization Problems. Ph.D. Dissertation. Massachusetts Institute of Technology, MA, U.S.A. (1988)
3. Bertsimas, D.J., Jaillet, P., Odoni, A.R.: A priori optimization. Oper. Res. **38**, 1019–1033 (1990)
4. Applegate, D.L., Bixby, R.E., Chvátal, V., Cook, W.J.: The Traveling Salesman Problem: A Computational Study. Princeton University Press, Princeton (2006)
5. Papadimitriou, C.H., Steiglitz, K.: Combinatorial Optimization: Algorithms and Complexity. Dover, New York (1998)
6. Li, W.: Constructing a solution attractor for the probabilistic traveling salesman problem through simulation. In: Emmerich, M., Deutz, A., Schuetze, O., Bäck, T., Tantar, E., Tantar, A.-A., Del Moral, P., Legrand, P., Bouvry, P., Coello, C.A. (eds.) EVOLVE – A Bridge between Probability, Set Oriented Numerics, and Evolutionary Computation. Advances in Intelligent Systems and Computing, vol. 227, pp. 1–15. Springer, Switzerland (2013)
7. Jaillet, P: Probabilistic Traveling Salesman Problems. Ph.D. Thesis, Massachusetts Institute of Technology, MA, U.S.A. (1985)
8. Jaillet, P.: A priori solution of a traveling salesman problem in which a random subset of the customers are visited. Oper. Res. **36**, 929–936 (1988)

9. Bertsimas, D.J., Howell, L.: Further results on the probabilistic traveling salesman problem. J. Oper. Res. **65**, 68–95 (1993)
10. Jézéquel, A.: Probabilistic Vehicle Routing Problems. Master thesis. Massachusetts Institute of Technology, MA, U.S.A. (1985)
11. Rossi, F., Gavioli, F.: Aspects of heuristic methods in the probabilistic traveling salesman problem. In: Advanced School on Stochastic in Combinatorial Optimization, pp. 214–227. World Scientific, Hackensack (1987)
12. Jaillet, P.: Analysis of probabilistic combinatorial optimization problems in Euclidean spaces. Math. Oper. Res. **18**, 51–70 (1993)
13. Birattari, M., Balaprakash, P., Stützle, T., Dorigo, M.: Estimation-based local search for stochastic combinatorial optimization using delta evaluations: a case study on the probabilistic traveling salesman problem. INFORMS J. Comput. **20**, 644–658 (2008)
14. Jaillet, P., Odoni, A.R.: The probabilistic vehicle routing problem. In: Golden, B.L., Assad, A.A. (eds.) Vehicle Routing: Methods and Studies, pp. 293–318. North-Holland, Amsterdam (1988)
15. Laporte, G., Louveaus, F., Mercure, H.: A priori optimization of the probabilistic traveling salesman problem. Oper. Res. **42**, 543–549 (1994)
16. Binachi, L., Knowles, J., Bowler, N.: Local search for the probabilistic traveling salesman problem: correction to the 2-p-opt and 1-shift algorithms. Eur. J. Oper. Res. **1611**, 206–219 (2005)
17. Bianchi, L., Campbell, A.M.: Extension of the 2-p-opt and 1-shift algorithm to the heterogeneous probabilistic traveling salesman problem. Eur. J. Oper. Res. **1761**, 131–144 (2007)
18. Liu, Y.-H.: Diversified local search strategy under scatter search framework for the probabilistic traveling salesman problem. Eur. J. Oper. Res. **191**, 332–346 (2008)
19. Weyland, D., Bianchi, L., Gambardella, L.M.: New approximation-based local search algorithms for the probabilistic traveling salesman problem. In: Moreno-Díaz, R., Pichler, F., Quesada-Arencibia, A. (eds.) EUROCAST 2009. LNCS, vol. 5717, pp. 681–688. Springer, Heidelberg (2009)
20. Marinakis, Y., Migdalas, M., Pardalos, P.M.: Expanding neighborhood search GRASP for the probabilistic traveling salesman problem. Optim. Lett. **23**, 351–360 (2008)
21. Cityplace Campbell, A.M.: Aggregation for the probabilistic traveling salesman problem. Comput. Oper. Res. **33**, 2703–2724 (2006)
22. Choi, J., Lee, J.H., Realff, M.J.: An algorithmic framework for improving heuristic solutions: part II, a new version of the stochastic traveling salesman problem. Comput. Chem. Eng. **28**(8), 1297–1307 (2004)
23. Bowler, N.E., Fink, T.M.A., Ball, R.C.: Characterization of the probabilistic traveling salesman problem. Phys. Rev. E **682**, 1–7 (2003)
24. Liu, Y.-H.: A hybrid scatter search for the probabilistic traveling salesman problem. Comput. Oper. Res. **34**, 2949–2963 (2007)
25. Liu, Y-H.: Solving the probabilistic traveling salesman problem based on genetic algorithm with queen selection scheme. In: Greco, F. (ed.) Traveling Salesman Problem, pp. 157–172. InTech (2008)
26. Liu, Y.-H., Jou, R.-C., Wang, C.-C., Chiu, C.-S.: An evolutionary algorithm with diversified crossover operator for the heterogeneous probabilistic TSP. In: Carbonell, J.G., Siekmann, J. (eds.) Modeling Decisions for Artificial Intelligence MDAI 2007, LNCS, vol. 4617, pp. 351–360. Springer, Berlin (2007)
27. Bianchi, L., Gambardella, L.M., Dorigo, M.: An ant colony optimization approach to the probabilitic traveling salesman problem. In: Guervós, J.J.M., Adamidis, P., Beyer, H.-G., Schwefel, H.-P., Fernández-Villacañas, J.-L. (eds.) Proceedings of the 7th International Conference on Parallel Problem Solving from Nature. LNCS, vol. 2439, pp. 883–892. Springer, London (2002)
28. Bianchi, L., Gambardella, L.M., Dorigo, M.: Solving the homogeneous probabilistic traveling salesman problem by the ACO metaheuristic. In: Dorigo, M., Di Caro, G., Sampels, M. (eds.) Proceedings of the 3rd International Workshop on Ant Algorithms. LNCS, vol. 2463, pp. 176–187. Springer, London (2002)

29. Gutjahr, W.J.: A conberging ACO algorithm for stochastic combinatorial optimization. In: Albrecht, A., Steinhöfel, K. (eds.) Proceedings of the 2nd Symposium on Stochastic Algorithms, Foundations and Applications. LNCS, vol. 2827, pp. 10–25. Springer, Berlin (2003)
30. Gutjahr, W.J.: S-ACO: an ant-based approach to combinatorial optimizaiton under uncertainty. In: Dorigo, M., Birattari, M., Blum, C., Gambardella, L.M., Mondada, F., Stützle, T. (eds.) Proceedings of the 4th International Workshop on Ant Colony Optimizaiton and Swarm Intelligence. LNCS, vol. 3172, pp. 238–249. Springer, Berlin (2004)
31. Branke, J., Guntsch, M.: Solving the probabilistic TSP with ant colony optimization. J. Math. Model. Algorithms **34**, 403–425 (2004)
32. Birattari, M., Balaprakash, P., Dorigo, M.: ACO/F-Race: ant colony optimization and racing techniques for combinatorial optimization under uncertainty. In: Doerner, K.F., Gendreau, M., Greistorfer, P., Gutjahr, W.J., Hartl, R.F., Reimann, M. (eds.) Proceedings of the 6th Metaheuristics International Conference, pp. 107–112 (2005)
33. Birattari, M., Balaprakash, P., Dorigo, M.: The ACO/F-race algorithm for combinatorial optimization under uncertainty. In: Doerner, K.F., Gendreau, M., Greistorfer, P., Gutjahr, W.J., Hartl, R.F., Reimann, M. (eds.) Metaheuristics – Progress in Complex Systems Optimization. Operations Research/Computer Science Interfaces Series, pp. 189–203. Springer, Berlin (2006)
34. Liu, Y-H.: A memetic algorithm for the probabilistic traveling salesman problem. In: IEEE Congress on Evolutionary Computation (CEC2008), pp. 146–152 (2008)
35. Marinakis, Y., Marinakis, M.: A hybrid multi-swarm particle swarm optimization algorithm for the probabilistic traveling salesman problem. Comput. Oper. Res. **37**, 432–442 (2010)
36. Homem-de-Mello, T.: Variable-sample methods for stochastic optimization. ACM Trans. Model. Comput. Simul. **132**, 108–133 (2003)
37. Kleywegt, A.J., Shapiro, A., Homen-de-Mello, T.: The sample average approximation method for stochastic discrete optimization. SIAM J. Optim. **12**, 479–502 (2001)
38. Bianchi, L.: Ant Colony Optimization and Local Search for the Probabilistic Traveling Salesman Problem: A Case Study in Stochastic Combinatorial Optimization. Ph.D. Dissertation, Universite Libre de Bruxelles, Brussels, Belgium (2006)
39. Shapiro, A.: A simulation-based approach to two-stage stochastic programming with recourse. Math. Prog. **813**, 301–325 (1998)
40. Verweij, B., Ahmed, S., Kleywegt, A.J., Nemhauser, G., Shapiro, A.: The sample average approximation method applied to stochastic routing problems: a computational study. Comput. Optim. Appl. **24**, 289–333 (2003)
41. Balaprakash, P., Pirattari, P., Stützle, T., Dorigo, M.: Adaptive sample size and importance sampling in estimation-based local searh for the probabilistic traveling salesman problem. Eur. J. Oper. Res. **199**, 98–110 (2009)
42. Balaprakash, P., Pirattari, P., Stützle, T., Dorigo, M.: Estimation-based metaheuristics of the probabilistic traveling salesman problem. Comput. Oper. Res. **37**, 1939–1951 (2010)
43. Tang, H., Miller-Hooks, E.: Approximate procedures of the probabilistic traveling salesperson problem. Transp. Res. Record **1882**, 27–36 (2004)
44. Li, W., Feng, M.: Solution attractor of local search in traveling salesman problem: concepts, construction and application. Int. J. Metaheuristics **23**, 201–233 (2013)
45. Li, W.: Seeking global edges for traveling salesman problem in multi-start search. J. Global Optim. **51**, 515–540 (2011)
46. Li, W.: A parallel multi-start search algorithm for dynamic traveling salesman problem. In: Pardalos, P.M., Rebennack, S. (eds.) 10th International Symposium on Experimental Algorithms, SEA2011. LNCS, vol. 6630, pp. 65–75. Springer, Heidelberg (2011)
47. Li, W., Feng, M.: A parallel procedure for dynamic multi-objective TSP. In: Proceedings of 10th IEEE International Symposium on Parallel and Distributed Processing with Applications, pp. 1–8. IEEE Computer Society (2012)
48. Sudman, S.: Applied Sampling. Academic Press, New York (1976)
49. Walson, J.: How to Determine a Sample Size. Peen Cooperative Extension, University Park, PA (2001)

50. Johnson, D.S., McGeoch, L.A.: The traveling salesman problem: a case study. In: Aarts, E., Lenstra, J.K. (eds.) Local Search in Combinatorial Optimization, pp. 215–310. Wiley, Chichester (1997)
51. Birattari, M., Balaprakash, P., Stützle, T., Dorigo, M.: Estimation-based Local Search for Stochastic Combinatorial Optimization. Technical report TR/IRIDIA/2007-003, IRIDIA, Université Libre de Bruxelles, Brussels, Belgium (2007)

Average Cuboid Volume as a Convergence Indicator and Selection Criterion for Multi-objective Biochemical Optimization

Susanne Rosenthal and Markus Borschbach

Abstract The performance of a multi-objective evolutionary algorithm (MOEA) is evaluated with regard to the quality of the populations under two aspects: the distance of the non-dominated set of a population to the true Pareto front (PF_{true}) and the spread among these solutions. Diverse convergence indicators have been proposed in the past with different requirements: either PF_{true} or a reference set of Pareto-optimal solutions is required. Furthermore, most of the convergence indicators are restricted to a non-dominated solution set, and therefore, the quality of the entire population is only represented by the non-dominated solutions. This work presents a statistically reasonable convergence indicator that is able to reflect the quality of the entire population. The average cuboid volume (ACV) assigns desirable aspects regarding the classification of entire populations. These preferable features are demonstrated and discussed. Furthermore, ACV is used as selection criterion to determine the solutions of the succeeding generation in a proposed customized NSGA-II for biochemical optimization. Two selection strategies based on the ACV indicator are proposed and empirically compared to a Pareto rank-based selection strategy. These selection strategies depend on two parameters and the adaption of the selection pressure by a variation of these parameters is empirically investigated on a three-dimensional biochemical minimization problem.

1 Introduction

The performance of MOEA is usually assessed by the quality of the non-dominated solutions in a population. A range of indicators have been developed in the past identifying different desirable aspects of the non-dominated solution sets by mapping these preferable properties into an unary value. The generally accepted desirable

S. Rosenthal (✉) · M. Borschbach
University of Applied Sciences, FHDW, Hauptstr. 2, 51465 Bergisch Gladbach, Germany
e-mail: susanne.rosenthal@fhdw.de

M. Borschbach
e-mail: markus.borschbach@fhdw.de

© Springer International Publishing AG 2017
M. Emmerich et al. (eds.), *EVOLVE – A Bridge Between Probability,*
Set Oriented Numerics and Evolutionary Computation VII,
Studies in Computational Intelligence 662, DOI 10.1007/978-3-319-49325-1_9

aspects are the distance of the achieved non-dominated solutions to PF_{true} as well as the uniform and wide spread of the non-dominated solutions among themselves. The distance of the non-dominated solutions is assessed by convergence indicators and the spread is measured by diversity metrics. In particular, the convergence indicators have requirements or disadvantages that make the use of these indicators impossible or inadequate: Several convergence indicators require the knowledge of PF_{true} or at least a Pareto-optimal solution set that are usually unknown in the case of real-world multi-objective optimization problems. Most of the convergence indicators assess the quality of a population only by the non-dominated solutions. This makes the identification of the entire population progress impossible. Other convergence indicators measure the quality of a solution set without referring to the set size. These indicators are inappropriate for comparison of populations with various sizes (e.g., [1, 2]).

This work presents a statistically reasonable convergence indicator that is able to evaluate the quality of the entire population, allows the comparison of populations of different sizes, and does not require the knowledge of PF_{true} or a Pareto-optimal set. This average cuboid volume (ACV) has been firstly introduced in [3]. It calculates the average volume of the cuboids spanned by the solutions of a population to a predefined ideal point.

Some of these state-of-the-art indicators have been used as selection criteria in a MOEA to assign high selection probabilities to high-quality solutions. The most established algorithm in this area is the SMS-EMOA introduced by Emmerich et al. that uses a hypervolume-based and steady-state selection strategy [4]. The solutions for the succeeding generations are selected by using the hypervolume on different subsets. The solution set with the maximum hypervolume is selected as succeeding generation. Another established algorithm is IBEA proposed by Zitzler and Künzli [5]. The optimization goal is defined in terms of an indicator that is used in the selection process of the MOEA. In general, every binary-quality indicator is usable as a selection criterion to determine the solutions of the succeeding generation. According to this work, the ACV indicator is used as a selection criterion to determine the solutions for the succeeding generation in a customized NSGA-II for biochemical optimization. Two ACV-based selection strategies are proposed and empirically compared to a Pareto rank-based selection strategy termed aggregate selection. This benchmarking is performed on a three-dimensional minimization problem and evaluated in terms of convergence and diversity. The experiments are evaluated with a focus on the following three questions:

1. Which selection strategy achieves the best performance with regard to the desired aspects convergence and diversity?
2. How does the variation of the selection parameters influence the performance of the customized NSGA-II?
3. How general are these results?

The remainder of this work is organized as follows: In the second section, we will give some definitions and a short review of the most popular convergence metrics. In Sect. 3, the convergence indicator ACV is presented and its properties are

demonstrated and discussed. In Sect. 4, the components of the customized NSGA-II
- first introduced in [6] and improved in [7, 8] - as well as the ACV-based selection
strategies are proposed. In Sect. 5, the simulation setups and the experiments are
presented and discussed. Section 6 provides a critical summary and gives answers to
the questions raised in this section.

2 Background and Review of Convergence Metrics

The following multi-objective minimization problem is considered

$$\min_{x \in Q}\{F(x)\}, \tag{1}$$

where Q is the decision (variable) space and F is defined as the objective vector
consisting of k objective functions $F : Q \longrightarrow \mathbb{R}$, $F(x) = (f_1(x), \ldots, f_k(x))$ with $f_i :
Q \longrightarrow \mathbb{R}$.

Definition 1

(a) A vector $u \in \mathbb{R}^n$ is said to dominate $v \in \mathbb{R}^n$ (denoted by $u \prec v$) if and only if
u is partially less than v: $\forall i \in \{1, \ldots, n\}$, $u_i \leq v_i$ and there exists at least one
$i \in \{1, \ldots, n\}$: $u_i < v_i$.

(b) A vector $u \in \mathbb{R}^n$ weakly dominates a vector $v \in \mathbb{R}^n$ (denoted by $u \preceq v$) if
$\forall i \in \{1, \ldots, n\}$: $u_i \leq v_i$.

(c) Consider a set of decision solutions $X \in \mathbb{R}^n$. The set X is termed a global Pareto-
optimal set if $\forall u \in X$, $\nexists v \in \mathbb{R}^n : v \prec u$.

(d) A point $r \in \mathbb{R}^n$ is further termed ideal point, if the coordinates r_i are simply
better than all feasible f_i.

Some metrics have been proposed to reflect the quality of a non-dominated solution
set in terms of convergence to PF_{true} with the final aim of evaluating the MOEA
performance. These metrics are evolved under the focus of convergence studies
or of some statistical consideration. An accurately interpretation of a metric value
regarding the relationship between two sets of non-dominated solutions requires a
theoretical analysis of the metric itself. Hansen and Jaszkiewicz [9] defined three
outperformance relations reflecting the relationship of two internally non-dominated
solution sets relative to PF_{true}.

Definition 2 A and B are internally non-dominated solution sets and $ND(S)$ denotes
the non-dominated solutions in the set S:

(a) **(Weak outperformance)**: $A \; O_w \; B \Leftrightarrow ND(A \cup B) = A$ and $A \neq B$. A weakly
   outperforms B if all solutions in B are covered (equal or dominated) by those in
   A and at least one solution in A is not contained in B.
(b) **(Strong outperformance)**: $A \; O_s \; B \Leftrightarrow ND(A \cup B) = A$ and $B \setminus ND(A \cup B) \neq \emptyset$.
   A strongly outperforms B if all solutions in B are covered (equal or dominated)
   by those in A and some solution in B is dominated by a solution in A.
(c) **(Complete outperformance)**: $A \; O_c \; B \Leftrightarrow ND(A \cup B) = A$ and $B \cap ND$
   $(A \cup B) = \emptyset$. A completely outperforms B if each solution in B is dominated
   by a solution in A.

A metric has to be compatible with these relations; otherwise, the metric values are
misleading. Therefore, Hansen and Jaszkiewicz further defined the compatibility or
weak compatibility with these outperformance relations:

Definition 3

(a) **Weak compatibility**: A metric is weakly compatible with an outperformance
   relation O if for two non-dominated solution sets A and B with $A \; O \; B$, such that
   the metric evaluates A as not being worse than B.
(b) **Compatibility**: A metric is weakly compatible with an outperformance relation
   O if for two non-dominated solution sets A and B with $A \; O \; B$, such that the
   metric evaluates A as being better than B.

The established metrics are reviewed in the following. A theoretical analysis of
several metrics according to the compatibility of the outperformance relations is
proposed in [10].

The hypervolume [11] or S-metric [12] is equivalent to the Lebesgue measure
[13] and determines the closeness of a non-dominated solution set to PF_{true} as well
as the spread of the overlapped dominated space. The more the solutions set approximates PF_{true}, the more the metric value increases, since this indicator is relative
to a predefined anti-optimal point. The hypervolume is one of the most established
metric because of its favorable mathematical properties [14]. One disadvantage of
this operator is the choice of the anti-optimal point as it influences the results and
is subject of ongoing research. Other disadvantages are its sensitivity to the relative
scaling of the objectives, the presence or the absence of extreme points of a front,
and the high computation complexity caused by the necessary point ordering $O(n^k)$
[15]. A lot of research has been done to find an implementation of the hypervolume
which reduces the computational complexity [16, 17].

Table 1 shows the computational complexity of different calculation methods for
the hypervolume. According to this overview, Walking Fish Group (WFG) is the
fastest calculation method for the exact hypervolume [18].

Another convergence metric is the D-metric introduced by Zitzler [12]. The starting point are two solution sets A, B. This metric calculates the size of the space
dominated by A and not dominated by B.

$$D(A, B) = H(A + B; r) - H(B; r),$$

Table 1 Overview of the different methods to calculate the exact hypervolume and the worst-case complexity, where n is the number of solutions and d the number of objectives

Methods	Worst-case complexity
HSO: Hypervolume by slicing objectives	$O(n^{d-1})$
FPL: Fonseca, Paquete and López-Ibánez	$O(n^{d-2} \cdot \log n)$
HOY: Hypervolume by Overmars and Yap	$O(n \cdot \log n + n^{d/2})$
WFG: Walking Fish Group	$O(2^n)$

where $H(A; r)$ denotes the hypervolume with the anti-optimal point r. A reference set is needed to assess the convergence to PF_{true}. Zitzler also proposed the C-metric [11] that is an appropriate measure to compare the dominance of two Pareto-optimal sets PF_1 and PF_2. The C-metric maps the ordered pair $(PF1, PF2)$ into the interval $[0; 1]$:

$$C(PF_1, PF_2) := \frac{|\{b \in PF_2 \mid \exists a \in PF_1 : a \preceq b\}|}{|PF_2|}$$

Therefore, the value $C(PF_1, PF_2) = 0$ means that no point of PF_2 weakly dominates at least one point of PF_1, whereas $C(PF_1, PF_2) = 1$ implicate that all points of PF_2 are weakly dominated by PF_1. This metric is usually not symmetric; therefore, $C(PF_1, PF_2)$ is not a metric in a mathematical sense and consequently $C(PF_1, PF_2)$ and $C(PF_2, PF_1)$ have to be determined.

The error ratio (ER) [19] is introduced by Veldhuizen that is a percentage measure for the number of solutions in a set that lies on PF_{true}. This metric requires PF_{true} as a reference set.

$$ER(PF_{approx}) = \frac{1}{|PF_{approx}|} \sum_{i=1}^{|PF_{approx}|} e_i \text{ whereas} \tag{2}$$

$$e_i = \begin{cases} 0 & \text{if the solution vector } i \text{ is in } PF_{approx} \\ 1 & \text{if the solution vector } i \text{ is not in } PF_{approx} \end{cases} \tag{3}$$

A measure of $ER \approx 1$ means that PF_{approx} comprises only a low number of solutions in PF_{true}, whereas a lower measure value indicates that many solutions are in PF_{true}. ER is exceptionally sensitive to the reference set PF_{true}: If a Pareto-optimal solution is not in PF_{true}, it is treated as non-optimal solution by ER. Furthermore, ER takes not the closeness of PF_{approx} to PF_{true} into account.

The 'generational distance' (GD) is also proposed by Veldhuizen [20]. This metric is a measure of the average distance between solutions of PF_{approx} and PF_{true} and is defined as:

$$GD(PF_{approx}) = \frac{\left(\sum_{i=1}^n d_i^p\right)^2}{n},$$

where n is the number of solutions in PF_{approx}, usually $p = 2$ and d_i is the Euclidean distance between each solution in PF_{approx} to its nearest located member on PF_{true}.

A value of $GD(PF_{approx}) = 0$ denotes that $PF_{approx} = PF_{true}$. However, the GD provides no information about homogeneity, spread, or dominance of PF_{approx} compared to PF_{true}.

The convergence metric proposed by Deb [21] determines the distance of PF_{approx} to a reference set of PF_{true}, further denoted as PF^*. $PF^* = \{a_1, a_2, \ldots, a_n\}$ is the solution set of the optimal Pareto front or the final approximate Pareto-optimal set obtained from a MOEA run. In each generation, the following steps have to be performed for the determination of this metric:

- Generate the non-dominated solution set $PF_{approx} = \{p_1, p_2, \ldots, p_n\}$.
- The smallest normalized Euclidean distance d_i for each solution of PF_{approx} to PF^* is calculated:

$$d_i = \min_{j=1,\ldots,n} \sqrt{\sum_{k=1}^{M} \left(\frac{f_k(a_i) - f_k(p_j)}{f_k^{max} - f_k^{min}} \right)^2 },$$

where M denotes the number of objective functions, f_k^{max} is the maximal and f_k^{min} is the minimal function value of the kth objective function of PF^*.

- The convergence metric value is calculated as the average normalized distance for all solutions in PF_{approx}

$$C(PF_{approx}) = \sum_{i=1}^{|PF_{approx}|} \frac{d_i}{|PF_{approx}|}$$

The lower the values for this metric, the better the convergence.

The averaged Hausdorff distance (Δ_p) as a performance measure is proposed by Schütze [22]. Δ_p is evolved from the model of GD [20] and the inverted general distance (IGD) [23] and is defined by:

$$\Delta_p(X_Y) = \max(GD(X, Y), IGD(X, Y))$$

$$= \max \left(\frac{1}{m} \left(\sum_{i=0}^{n} dist(x_i, Y)^p \right)^{1/p}, \frac{1}{n} \left(\sum_{i=0}^{n} dist(y_i, X)^p \right)^{1/p} \right)$$

with the finite non-empty sets $X = \{x_1, x_2, \ldots, x_n\}$ and $Y = \{y_1, y_2, \ldots, y_n\}$, where X is regarded as the Pareto-optimal set and $Y = PF_{true}$.

Trautmann [24] recently proposed the R2 indicator that evaluates the quality of PF_{approx} regarding the convergence to PF_{true}, the solutions spread and the representation of the Pareto front shape. The R2 indicator is defined by:

$$R2(S, W, r) = \frac{1}{N} \sum_{w \in W} \min_{s \in S} \max_{j} \left(w_j \cdot (s_j - r_j) \right),$$ (4)

where $W = \{w^1, \ldots, w^N\} \subset \mathbb{R}^k$ is a set of N weight vector, $S \subset \mathbb{R}^k$ a set of solutions and $r \in \mathbb{R}^k$ is an ideal point that usually is chosen as an optimal objective vector better than all feasible solutions.

This indicator is popular for its computational complexity $O(Nk \cdot |S|)$ indicating that the complexity is linear with the number of weights, the problem dimension, and the solution set size. The number and the choice of weight vectors is an open issue, especially for $k > 2$. The volume of the space is exponentially increasing with k and potentially also the number of weight vectors. This makes the calculation of R2 as expensive as the hypervolume from a specific number of k on [25].

Empirical results have shown that the R2 indicator and the hypervolume are correlated by the Pearson's correlation coefficient with a statistically significant values of 0.76 [26].

3 Introduction of the Average Cuboid Volume

3.1 Motivation for the Average Cuboid Volume

The reason for the development of a new convergence metric is multiple: The disadvantage of the metrics D-metric, ER, GD, Δ_p, and the convergence metric of Deb is their dependency on the knowledge of PF_{true} or at least a reference set of Pareto-optimal solutions that are usually unknown in the case of real-world multi-objective problems (MOPs). Further, these metrics are not useful indicators for an entire ranking between generations of different sizes. However, the populations in MOGA are generally bounded in size. From a more global point of view, the evaluation and comparison of the global convergence behavior of whole populations - not only the non-dominated solution set of a generation - with respect to the influence of the population size or the selection pressure are required. For this purpose, a new metric has been developed with the definite goal of comparing the convergence behavior of whole populations of different sizes in a statistically meaningful way. Therefore, it is a 'fair' indicator for comparing generations of different sizes. This average cuboid volume (ACV) is developed according to the model of the hypervolume. The motivation for the exploitation of the hypervolume model is to profit from its preferable properties as mentioned above. The benefit of this new metric compared to the hypervolume is the low computational complexity as no point ordering is required. The computational complexity of ACV for a solution set of n individuals and k objective is $O(n \cdot k)$. The metric calculates the average cuboid volume of the cuboids spanned by the solution points with respect to a predefined ideal point r that is defined in Sect. 2. This ideal point is chosen as a theoretical optimal point of (1) and not as an anti-optimal one like for the hypervolume. In many MOPs or 'black box' optimization problems, it is easier to find an optimal point than an anti-optimal one - especially in the case of the three-dimensional biochemical minimization problem presented in the next section. ACV is calculated by

$$ACV(X) = \frac{1}{n} \sum_{i=1}^{n} \left(\prod_{j=1}^{k} (x_{ij} - r_j) \right), \qquad (5)$$

where n is the population size, k is the number of objectives, x_i are the solutions on the population X, and x_{ij} is the *jth* component of a solution x_i. By Definition 1, we have $x_{ij} - r_j \geq 0$. The lower the indicator values, the better is the global convergence behavior as the ideal point is chosen as a theoretical optimal point.

3.2 Discussion of the Averaged Cuboid Volume

The question according to the suitability of a metric for evaluation depends on the intention of the investigation object and the preferences. ACV is intended to evaluate the global convergence behavior of a whole population with the ultimate aim of comparing solution sets of different sizes according to the proximity to PF_{true}.

The first reason which is important for the use of ACV is that the convergence quality does not change in the case of multiple copies of one solution. ACV does not fulfill this averaging strategy that can be manifested through the following example: Let $x \in \mathbb{R}^k$ be a solution for Eq. (1). Further, $Y = \{x, x, \ldots, x\}$ is a set containing n copies of the solution x, then

$$ACV(Y) = \frac{1}{n} \sum_{i=0}^{n} \left(\prod_{j=1}^{k} (x_j - r_j) \right) = \frac{1}{n} \cdot n \prod_{j=1}^{k} (x_j - r_j) = \prod_{j=1}^{k} (x_j - r_j) = ACV(X)$$

The second reason is due the following observation: An intuitive indicator reflecting the quality of approximation sets of different Pareto front refinements requires 'better' indicator values for the finest approximation set. The following example demonstrates this effect for ACV:

Example 1 The Pareto front is given by the bounded convex function $f(x) = 1/x^2$ between the points $y_1 = (0.1, 100)$ and $y_2 = (1.1, 0.826)$ meaning

$$PF_{true} = \{(x, y) | y = 1/x^2 \text{ with } x \in [0.1, 1.1]\}. \qquad (6)$$

We consider the following three approximation sets of increasing refinement of the Pareto front

Table 2 Indicator values of ACV for the approximation sets $Y_1 - Y_3$ with the ideal point $(0, 0)$

X	Y_1	Y_2	Y_3
ACV(X)	3.13	2.75	2.43

$$Y_1 = \{(0.1 + 0.2 \cdot i, 1/(0.1 + 0.2 \cdot i)^2) \mid i \in \{0, 1, \dots, 5\}\},$$
$$Y_2 = \{(0.1 + 0.1 \cdot i, 1/(0.1 + 0.1 \cdot i)^2) \mid i \in \{0, 1, \dots, 10\}\},$$
$$Y_3 = \{(0.1 + 0.01 \cdot i, 1/(0.1 + 0.01 \cdot i)^2) \mid i \in \{0, 1, \dots, 100\}\}.$$

Table 2 depicts the indicator values of ACV for the three approximation sets with the ideal point $(0, 0)$.

The third reason of this indicator is the averaging effect. It is trivial that a dominating solution x yields better indicator values than the dominated one y, because $ACV(\{x\}) = \prod_{i=1}^{k}(x_j - r_j) < \prod_{i=1}^{k}(y_j - r_j) = ACV(\{y\})$. Moreover, if a dominated solution x_1 in the set $X = \{x_1, x_2, \dots, x_n\}$ is replaced by a dominating one $\bar{x}_1$, then $ACV(\{x_1, x_2, \dots, x_n\}) > ACV(\{\bar{x}_1, x_2, \dots, x_n\})$. The averaging effect is illustrated by the following example [21]:

Example 2 The true discrete Pareto front is described by $P = \{p_i | p_i = (0.1 \cdot (i - 1), 1 - (i - 1) \cdot 0.1)$ with $i = 1, \dots, 11\}$. Two solution sets are given by $X_1 = \{x_{1,1}, p_2, \dots, p_{11}\}$ and $X_2 = \{x_{2,1}, x_{2,2}, \dots, x_{2,11}\}$ with the elements $x_{1,1} = (\epsilon, 10)$ and $x_{2,i} = p_i + (\frac{\epsilon}{2}, 5)$ with $i = 1, \dots, 11$. For the outlier $x_{1,1}$, the values $\epsilon = 0.001$ are used for numerical evaluations. X_1 is a better approximation of the true Pareto front than X_2 as all solutions exceeding the outlier $x_{1,1}$ are positioned on the Pareto front. All points of X_2 are shifted by $(\frac{\epsilon}{2}, 5)$ from the Pareto front, but the difference of each element to PF_{true} is less than the outlier $x_{1,1}$. As we are interested in an averaging effect, the indicator values of X_1 have to be better than the one of X_2. This is true for $ACV(X)$ as $ACV(X_1) = 0.15$ and $ACV(X_2) = 2.65$ with the ideal point $(0, 0)$.

This indicates that ACV fulfills the important complement property of location parameters [27]. The complement property is formulated as an axiom:

Axiom 1 Given are n values $x_1, x_2, \dots, x_n$ with the location parameter M_n. In the case that a further value x_{n+1} enters the set, the following statements hold for the new location parameter $M(\{x_1, x_2, \dots, x_{n+1}\}) = M_{n+1}$:
  if $x_{n+1} \geq M_n$, then $M_{n+1} \geq M_n$;
  if $x_{n+1} \leq M_n$, then $M_{n+1} \leq M_n$

The complement property is important for the robustness of a measure and this property is further proven for the ACV indicator regarding the comparison of two solution sets:

Proposition 1 *Given are two solution sets* $X = \{x_1, \ldots, x_n\}$ *and* $Y = \{y_1, \ldots, y_{m+l}\}$ *with* $m, n, l \in \mathbb{N}$ *and it holds:*

    (i) $\forall i \in \{1, \ldots, n\}, \forall j \in \{1, \ldots, m\}: y_j \preceq x_i$ *and*

    (ii) $\forall i \in \{1, \ldots, n\}, \forall j \in \{m+1, \ldots, m+l\}: y_j \prec x_i$

Then, $ACV(Y) < ACV(X)$.

Proof It has to be shown that

$$ACV(X) > ACV(Y) \Leftrightarrow \frac{1}{n} \sum_{i=1}^{n} \left(\prod_{j=1}^{k} (x_{ij} - r_j) \right) > \frac{1}{m+l} \sum_{i=1}^{m+l} \left(\prod_{j=1}^{k} (y_{ij} - r_j) \right)$$

$$\Leftrightarrow (m+l) \sum_{i=1}^{n} \left(\prod_{j=1}^{k} (x_{ij} - r_j) \right) > n \sum_{i=1}^{m+l} \left(\prod_{j=1}^{k} (y_{ij} - r_j) \right) \quad (7)$$

It holds,

$$(m+l) \sum_{i=1}^{n} \left(\prod_{j=1}^{k} (x_{ij} - r_j) \right) \geq (m+l) \cdot n \cdot \min_{i=1,\ldots,n} \left(\prod_{j=1}^{k} (x_{ij} - r_j) \right)$$

According to the conditions (i) and (ii), it holds

$$(m+l) \min_{i=1,\ldots,n} \left(\prod_{j=1}^{k} (x_{ij} - r_j) \right) \geq (m+l) \max_{i=1,\ldots,m+l} \left(\prod_{j=1}^{k} (y_{ij} - r_j) \right) > \sum_{i=1}^{m+l} \left(\prod_{j=1}^{k} (y_{ij} - r_j) \right)$$

From these inequalities, Eq. (7) is proven:

$$(m+l) \sum_{i=1}^{n} \left(\prod_{j=1}^{k} (x_{ij} - r_j) \right) \geq (m+l) \cdot n \cdot \min_{i=1,\ldots,n} \left(\prod_{j=1}^{k} (x_{ij} - r_j) \right) > n \cdot \sum_{i=1}^{m+l} \left(\prod_{j=1}^{k} (y_{ij} - r_j) \right)$$

q.e.d.

This proposition implies that ACV is strictly monotonic to the Pareto compliance [28]: If solution set A is strictly better than a solution set B, then the indicator value of A ($I(A)$) is also strictly better than the one of B ($I(B)$): $A \preceq B \wedge B \not\preceq A \Rightarrow I(A) < I(B)$. The conditions (*ii*) and (*iii*) of the proposition imply that Y is strictly better than X and the indicator monotonicity is proven by Proposition 1.

    Finally, the compatibility of ACV with the outperformance relations is analyzed according to Definitions 2 and 3.

Proposition 2 *ACV is compatible with the complete and the strong outperformance relation, but not compatible with the weak outperformance relation.*

Proof First, the compatibility with the complete outperformance relation is proven: It has to be shown that if the set $A = \{x_1, x_2, \ldots, x_m\}$ completely outperforms the set

$B = \{y_1, y_2, \ldots, y_n\}$ with $m, n \in \mathbb{N}$ and ACV is compatible with the outperformance relation, then $x_i \prec y_j, \forall i, j \in \mathbb{N} \Rightarrow ACV(A) < ACV(B)$.

$ACV(B) > ACV(A)$ is proven by the following estimation under the conditions $x_i \prec y_j, \forall i, j$ and Axiom 1:

$$ACV(B) = \frac{1}{n} \cdot \sum_{i=1}^{n} \left(\prod_{j=1}^{k} (y_{ij} - r_j) \right) \geq \frac{1}{n} \cdot n \cdot \min_{i=1,\ldots,n} \left(\prod_{j=1}^{k} (y_{ij} - r_j) \right)$$

$$> \max_{x_i \prec y_j \; i=1,\ldots,m} \left(\prod_{j=1}^{k} (x_{ij} - r_j) \right) \underset{Ax.1}{\geq} \frac{1}{m} \cdot \sum_{i=1}^{m} \left(\prod_{j=1}^{k} (x_{ij} - r_j) \right) = ACV(A)$$

q.e.d.

Second, the compatibility of ACV with the strong outperformance is proven: The set $A = \{x_1, x_2, \ldots, x_m\}$ strongly outperforms set $B = \{y_1, y_2, \ldots, y_{n+l}\}$ with $m, n, l \in \mathbb{N}$ if

$$x_i \preceq y_j, \quad \forall i,j \quad \text{and} \quad \exists p \in \{1, \ldots, m\}, \quad \forall j \in \{n+1, \ldots, n+l\}: x_p \prec y_j \quad (8)$$

Under these conditions, it has to prove that $ACV(A) < ACV(B)$:

$$ACV(B) = \frac{1}{n+l} \sum_{i=1}^{n+l} \left(\prod_{j=1}^{k} (y_{ij} - r_j) \right)$$

$$= \frac{1}{n+l} \left(\sum_{i=1}^{n} \left(\prod_{j=1}^{k} (y_{ij} - r_j) \right) + \sum_{i=n+1}^{n+l} \left(\prod_{j=1}^{k} (y_{ij} - r_j) \right) \right)$$

$$\geq \frac{1}{n+l} \left(n \cdot \min_{i \in 1,\ldots,n} \prod_{j=1}^{k} (y_{ij} - r_j) + l \cdot \min_{i=n+1,\ldots,n+l} \prod_{j=1}^{k} (y_{ij} - r_j) \right)$$

$$> \frac{1}{n+l} \left(n \cdot \max_{i \in 1,\ldots,n} \prod_{j=1}^{k} (x_{ij} - r_j) + l \cdot \prod_{j=1}^{k} (x_{pj} - r_j) \right)$$

$$> \frac{1}{n} \cdot (n+l) \cdot \max \left(\max_{i \in 1,\ldots,n} \prod_{j=1}^{k} (x_{ij} - r_j), \prod_{j=1}^{k} (x_{pj} - r_j) \right)$$

$$\underset{Ax.1}{\geq} \frac{1}{m} \sum_{i=1}^{m} \left(\prod_{j=1}^{k} (x_{ij} - r_j) \right) = ACV(A)$$

q.e.d.

Third, the compatibility of ACV with the weak outperformance relation is disproven by the following example: The set $A = \{(0.2, 0.6), (0.3, 0.4)\}$ weakly outperforms the set $B = \{(0.2, 0.6), (0.3, 0.4), (0.7, 0.2)\}$, but $ACV(A) = 0.126$ and $ACV(B) = 0.12$, and therefore, $ACV(A) > ACV(B)$.

The use of $ACV(X)$ as a convergence metric and as a diversity metric is not part of our requirements. $ACV(X)$ is not a reliable indicator for diversity. A solution set with clustered solutions does not always achieve better indicator values, which is demonstrated in the following example:

Example 3 Once more PF_{true} is described by Eq. (3) and the solution set $Y_4 = \{(0.29, 11.89), (0.3, 11.11), (0.31, 10.4), (0.32, 9.77), (0.33, 9.18), (0.34, 8.65)\}$ contains clustered solutions on the true Pareto front, then $ACV(Y_4) = 3.18 \approx ACV(Y_1)$. Though the solutions of Y_4 are much more clustered than those of Y_1 and Y_4 receive nearly the same indicator values as Y_1.

Furthermore, the use of the ACV indicator as a selection criterion results in very clustered solutions on one part of the Pareto front and makes a further diversity preserving method necessary. This effect is demonstrated by the following simple example.

Example 4 Two Pareto fronts are given by the bounded convex functions $f(x) = 2/x$ and $g(x) = 1/x^2$ between the x-coordinates 0.1 and 1.1, meaning

$$PF1_{true} = \{(x, y)|y = 2/x \text{ with } x \in [0.1, 1.1]\} \tag{9}$$

$$PF2_{true} = \{(x, y)|y = 1/x^2 \text{ with } x \in [0.1, 1.1]\} \tag{10}$$

We consider an approximation set for each Pareto front: X_1 is an approximation set for $PF1_{true}$ and X_2 is one for $PF2_{true}$. The solutions are each positioned at the boundaries of the Pareto fronts (Fig. 1).

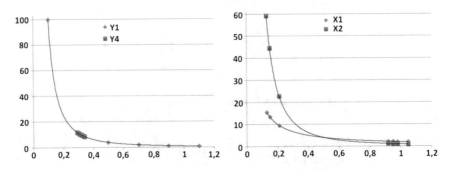

Fig. 1 Visualization of Example 3 (*left figure*) and Example 4 (*right figure*)

Table 3 ACV value for each solution of the approximation sets with the ideal point $(0, 0)$

ACV(X)	x_1	x_2	x_3	x_4	x_5	x_6	x_7
X_1	2	2	2	2	2	2	2
X_2	7.7	6.66	4.76	1.1	1.05	1.02	0.95

Table 4 Overview of the properties fulfilled or incompatible with the ACV indicator

1	Insensitivity for multiple copies of equal solutions
2	Sensitivity for refined Pareto-optimal solution sets
3	Compatible with the Pareto compliance of Zitzler
4	Compatibility with the strong and complete outperformance relation
5	Not compatible with the weak outperformance *relation*

$$X_1 = \{x_1(0.13, 15.38), x_2(0.15, 13.33), x_3(0.21, 9.52), x_4(0.92, 2.17), x_5(0.95, 2.11),$$
$$x_6(0.98, 2, 04), x_7(1.05, 1.91)\},$$
$$X_2 = \{x_1(0.13, 59.17), x_2(0.15, 44.44), x_3(0.21, 22.68), x_4(0.92, 1.18), x_5(0.95, 1.11),$$
$$x_6(0.98, 1.04), x_7(1.05, 0.91)\},$$

Table 3 depicts the indicator values of each solution in the approximation sets determined with the ideal point $(0, 0)$.

In the case of the approximation set X_1, all solutions have the same probability to be chosen for reproduction, whereas in the case of X_2, the solutions at the right boundary are preferred by the selection strategy based on the ACV indicator. In conclusion, if the solutions on the Pareto front do not exhibit the same ACV values, the search process is guided in the direction of the lowest ACV values and therefore results in clustered solutions on one part of the Pareto front.

Summarizing the following Table 4 gives an overview of the properties that are fulfilled or incompatible with the ACV indicator.

4 The Components of the Customized NSGA-II

The customized NSGA-II differs to the traditional NSGA-II [29] in the components variation operators and selection strategy to determine the succeeding generation. The components variation operators, the encoding as well as the fitness functions constituting the three-dimensional biochemical minimization problem are described in the following.

4.1 Encoding

The individuals are implemented as 20-character strings symbolizing short peptide sequences of the length 20. These individuals are composed of 20 different characters representing the 20 canonical amino acids. This encoding is highly intuitive in the way that no transformation is required for fitness evaluation. Furthermore, it represents all feasible and most important - only feasible - solutions. Small changes performed by a variation operator on the character strings preserve the similarity of the created offsprings to their parents. A peptide in the customized NSGA-II is represented by

Example 5 Peptide: AACMNKKLSTRAAEEGGGTT.

The commonly bit string presentation is inadequate as redundancies are highly probabilistic and every peptide has therefore a different probability to be produced. Furthermore, the string encoding schemata also represents non-feasible peptides.

4.2 Variation Operators

Several mutation and recombination operators have been developed and are tested within the customized NSGA-II [6–8]. The combination of recombination and mutation operator which achieved the best convergence–diversity balance is the linear recombination operator 'LiDeRP' [7] and the adaption of the deterministic dynamic mutation of Bäck and Schütz [6]. The combination of LiDeRP and the deterministic dynamic mutation provides the most successful balance of exploitation and exploration of the search process:

The recombination operator LiDeRP varies the number of recombination points within the generations via a linearly decreasing function:

$$x_R(t) = \frac{l}{2} - \frac{l/2}{T} \cdot t, \tag{11}$$

which depends on the length of the individual l, the total number of the GA generations T and the index of the actual generation t. Three individuals are randomly selected from the current population for reproduction. The number of parents have been empirically verified in [30].

The deterministic dynamic operator of Bäck and Schütz [31] determines the mutation probabilities via the following function with $a = 2$

$$p_{BS} = \left(a + \frac{l-2}{T-1} t \right)^{-1}, \tag{12}$$

The mutation rate is bounded by $(0; \frac{1}{2}]$. As a high mutation rate in the early generations results in an inappropriate high destruction of the sequence structure and the probability that an optimal partial structure is lost by the populations increases, the start mutation rate is reduced by $a = 5$.

4.3 Fitness Functions

A three-dimensional biochemical minimization problem is used for benchmarking. The biochemical optimization problem has been constituted to be as generic as possible regarding the approximation technique to predict biochemical features by the primary structure of a peptide. The biochemical objective values are calculated by the descriptor values of the amino acids composing the peptide sequence.

The Needleman–Wunsch algorithm is one fitness function calculating the sequence alignment score of a peptide to a predefined reference peptide. This score indicates the structure similarity of these two peptides. The selection of this algorithm as a fitness function is motivated by the principle that molecular structure similarity is often related to similar molecule properties. The Needleman–Wunsch algorithm is implemented by the BioJava library [32].

The second fitness function is the determination of the molecular weight. A low molecular weight is a very important feature in the field of peptide-based drug design as a peptide with a low molecular weight ensures a better diffusion of the drug through the epithelial layer [33]. This function is also provided by the BioJava library. The molecular weight is computed as the sum of mass of each amino acid plus a water molecule: $\sum_{i=1}^{l} mass(a_i) + 17.0073(OH) + 1.0079(H)$, according to the periodic system of elements: oxygen (O) and hydrogen (H).

The third fitness function is the average hydrophilicity value of a peptide determined by the hydrophilicity scale of Hopp and Woods and a window size according to the peptide length [34]. The average hydrophilicity is calculated by: $\frac{1}{l} \cdot (\sum_{i=1}^{l} hydro(a_i))$. The hydrophilicity is an important feature of a peptide-based drug as the hydrophilic character is essential for the ability to cross cell membranes [35].

Also the objective values of the fitness functions molecular weight and average hydrophilicity are calculated as the absolute value of the difference between the objective function value of a candidate solution and the corresponding value of the predefined reference peptide.

4.4 Selection Strategy Used in the Customized NSGA-II

This section motivates the development of three selection strategies and describes its procedures. These strategies determine the individuals of the succeeding generations in the customized NSGA-II and replace the environment selection of the traditional NSGA-II which is based on the crowding distance and the Pareto ranking.

4.4.1 Motivation of the Selection Strategies

Each of the three selection procedures incorporate the three fundamental objectives of an ideal population constitution achieved by an evolutionary process in the area of molecule optimization:

- Diversity of the genetic material: This objective refers to the creation of an at most high diversity of the genetic material within the succeeding generation.
- Maximization of the solutions spread: This objective refers to the detection of high-quality individuals with an at most wide spread among these individuals.
- High fitness-directional guidance: This objective refers to the strong guidance of the selection process in the direction of high-quality individuals.

Another important aspect of the selection process especially in the field of molecule optimization is the role of change in the selection procedure. The proposed three selection procedures represent, respectively, an aggregation of the three selection objectives and the aspect of change: Two selection strategies are based on the tournament selection to integrate the aspect of change in the selection strategies and make use of a combination of fitness-proportionate selection and a discerning selection criterion. The third strategy uses random selection instead of the fitness-proportionate selection.

The three selection strategies depend on two parameters, a probability parameter controlling the strength of the fitness-directional guidance and the tournament size. The interdependence of these two parameters on the selection procedure is empirically verified in the experiments on the generic three-dimensional biochemical minimization problem. In general, these selection strategies are characterized as preservative procedures as each individual has a nonzero change to be selected for the succeeding generation.

4.4.2 Aggregate Selection

The first selection strategy is tournament-based and uses a combination of front-based stochastic universal sampling (SUS) and a rank-based discerning selection criterion. The procedure is depicted in Fig. 2 and is denoted as aggregate selection referring to the aggregation of the three selection objectives. The procedure of the aggregate selection strategy starts with tournament selection of the size TS. This tournament set is Pareto ranked. An individual is chosen from the Pareto-optimal front with a probability p_0. With a probability of $1 - p_0$, individuals are chosen from the different fronts via SUS. The number N of pointers refers to the number of fronts and the segments are equal in size to the number of individuals in each front. Therefore, the parameters of this selection strategy are the TS and p_0.

Front-based SUS ensures the diversity of the genetic material and a potentially high solutions' spread. Further, it provides the opportunity to low-quality solutions to find their way in the succeeding generation. Low-quality solutions potentially have high-quality genetic motifs, which produce high-quality solutions in later generations

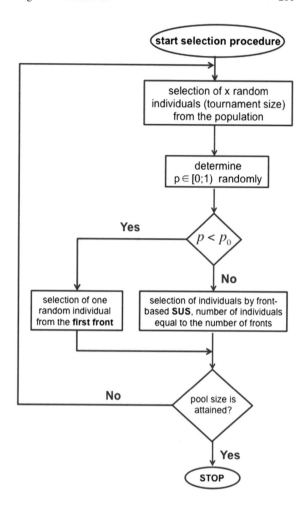

Fig. 2 Aggregate selection strategy

and ensure genetic diversity. The selection of an individual from the Pareto-optimal front is characterized as the rank-based discerning selection criterion and ensures the fitness-directional guidance.

4.4.3 ACV-Based and ACV-Random Selection Strategy

The procedure of the ACV-based selection strategy is comparable to the aggregate selection in its procedure (Fig. 3), whereas the rank-based discerning selection criterion is substituted by a discerning selection criterion using the ACV indicator: The ACV_{scaled} value for each individual in the tournament set is determined via Eq. (13) with $X = \{x_0\}$ and the individual with the lowest ACV_{scaled} value is selected. The part of the process differing from aggregate selection is highlighted. The basic idea

Fig. 3 ACV-based selection
strategy with SUS

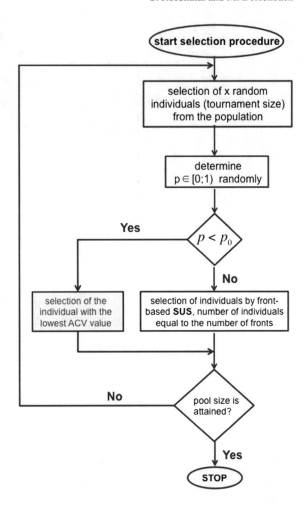

for the ACV-based selection criterion is motivated by the following consideration of
the aggregate selection strategy: The randomly chosen individuals by the tournament
selection are ranked, and a random individual from the first front is selected. This
selection from the first front does not guarantee the selection of the fittest individual
with respect to all objective values as the Pareto ranking is relative to the other individuals in the tournament set. This criterion is substituted by the determination of
an ACV_{scaled} value for each individual of the tournament set. In the case of multiple
lowest ACV values, a random one is selected. This criterion guarantees the selection
of the fittest individual with respect to all objectives within the tournament set.

The motivation for the ACV-random selection is the empirical investigation of
the influence of the front-based SUS on the search process. Therefore, this part is
replaced by a simple random selection of an individual from the tournament set with

Fig. 4 ACV-random
selection strategy

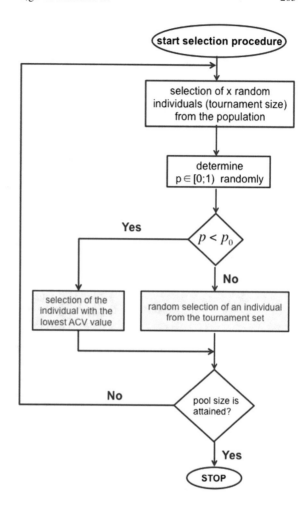

a probability of $1 - p_0$. The procedure is depicted in Fig. 4 and the differing selection criterions compared to the aggregate selection are highlighted. The parameters of the ACV-based and the ACV-random selection are TS and p_0.

5 Simulation Setups and Experiments

5.1 Simulation Setups

Two series of tests are performed: The first series of tests compares the performances of the three selection strategies with different p_0 settings regarding convergence, diversity, and relative non-dominated solution quality. The probability

$p_0 = 0\%$ denotes pure SUS as a selection strategy. The start population has a size
of 100 randomly initialized individuals representing 20-mer peptides. In the second
series of test, the influence of the tournament size on the performance is empirically
investigated.

For statistical reasons, each configuration is run 30 times until the 18th generation
as we are focused on early convergence [6, 7]. A normalized version of the ACV
indicator is used as convergence metric which ensures that all objective function
values have the same influence on the indicator values:

$$ACV_{scaled} = \frac{1}{n} \sum_{i=1}^{n} \left(\prod_{j=1}^{k} \frac{(x_{ij} - r_j)}{\bar{x}_j} \right), \text{ with } \bar{x}_j = max_i\{x_{ij}\}, \forall j = 1, \ldots, k \quad (13)$$

A relative ACV measure is used as relative non-dominated solution quality to evaluate
the average cuboid volume by the solutions of the non-dominated solutions in relation
to the average cuboid volume of the entire population:

$$ACV_{rel} = \frac{\frac{1}{f} \sum_{i=1}^{f} (\prod_{j=1}^{k} (x_{ij} - r_j))}{\frac{1}{n} \sum_{i=1}^{n} (\prod_{j=1}^{k} (x_{ij} - r_j))}, \quad (14)$$

The ideal point is generally chosen as $(0, 0, 0)$. The diversity within the population
is assed via:

$$\Delta = \sum_{i,j=1,i<j} \frac{|d_{ij} - \bar{d}|}{N} \quad \text{with } N = \binom{n}{2} = \frac{n(n-1)}{2},$$

where d_{ij} is the Euclidean distance of each possible combination of solutions, n is
the number of solutions and $\bar{d}$ is the average distance of all distances. Box plots
are created of the three performance indicators for each configuration. The indicator
values are scaled for an optimal graphical representation.

5.2 Evaluation

A good performance is achieved by a good convergence–diversity balance and by
an at most low value of the relative non-dominated solution quality of a configuration. A good convergence–diversity balance is given by an at most low value of
ACV_{scaled} and an at most high diversity values. Numerical indicators are used to support the following statements. The results of the first series of tests are represented in
Figs. 5, 6, and 7. All three selection strategies reveal a decrease of the ACV_{scaled} values by an increase of the probability p_0 for the fitness-directional guidance (Fig. 5).
The increase of p_0 results in a decrease of the diversity in the case of the aggregate selection strategies and ACV-random selection (Fig. 6). The diversity of the

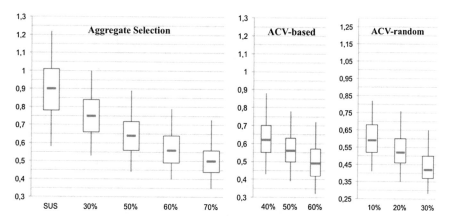

Fig. 5 ACV_{scaled} of the three selection strategies with different p_0-values and $TS = 10$

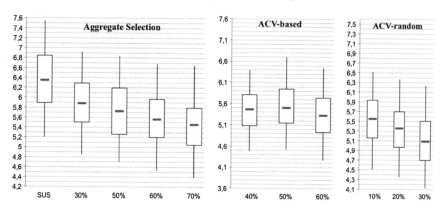

Fig. 6 Diversity of the three selection strategies with different p_0-values and $TS = 10$

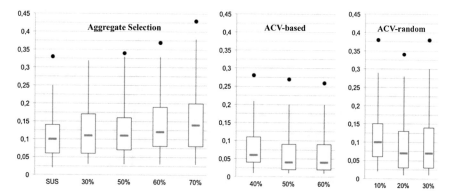

Fig. 7 ACV_{rel} of the three selection strategies with different p_0-values and $TS = 10$

ACV-based selection strategy is on a similar level for all three p_0 settings. The relative non-dominated solution quality measured by ACV_{rel} reveals a general tendency to lower ACV_{rel} values and therefore a tendency to a higher non-dominated solution quality relative to the quality of the entire population in the case of ACV-based and ACV-random selection. The ACV_{rel} results of ACV-random selection are generally higher than those of ACV-based selection. However, the ACV_{rel} results of the aggregate selection indicate an increase of the relative non-dominated solution quality. This confirms the hypothesis that the selection of a random individual of the Pareto-optimal front in the case of aggregate selection does not guarantee the selection of the fittest individual with respect to all objective values. The ACV indicator is an adequate measure for this purpose. According to [36], a good convergence–diversity balance is given by $p_0 = 60\%$ in the case of aggregate selection and $p_0 = 50\%$ for ACV-based selection. As the convergence and diversity results of ACV-random selection with $p_0 = 10\%$ are comparable to the results of Aggregate and ACV-based selection with the optimal settings, $p_0 = 10\%$ is regarded as advisable settings for $ACV - random$ selection.

The results of the second series of tests are represented in Figs. 5, 6, and 7. In these configurations, the three selection strategies are used with the corresponding optimal parameter setting p_0. The ACV_{scaled} results of aggregate selection are decreased for $TS = 10$ and higher values and reveal a convergence improvement (Fig. 8). The variation of TS reveals no influence on the diversity values. The relative non-dominated solution quality is increased for $ts = 10$ and 12. ACV-based as well as ACV-random selection result in comparable ACV_{scaled} values independent of the TS settings (Figs. 9, 10). The ACV_{rel} values are decreased for $TS = 10$ and higher revealing a higher relative non-dominated solution quality. The diversity is significantly increased for $TS = 10$ and higher in the case of the ACV-based selection. A tendency to higher diversity is also observable for $TS = 10$ and higher in the case of ACV-random selection.

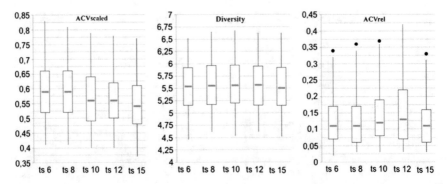

Fig. 8 Performance of configuration with aggregate selection, $p_0 = 60\%$ and a variation of TS

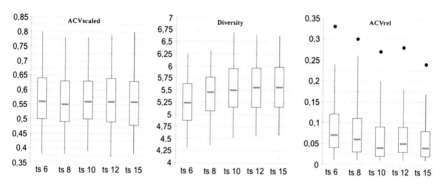

Fig. 9 Performance of configuration with ACV-based selection, $p_0 = 50\%$ and a variation of *TS*

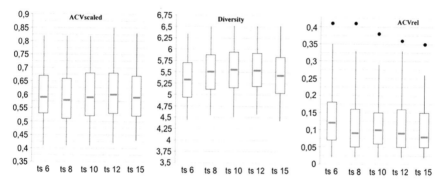

Fig. 10 Performance of configuration with ACV-random selection, $p_0 = 10\%$ and a variation of *TS*

6 Conclusion

The evaluation of a MOEA performance is based on two basic principles: The distance of the non-dominated solutions to the true Pareto front and the spread among the solutions. Diverse convergence metrics have been proposed in the literature with several disadvantages: Some convergence metrics require the knowledge of the Pareto-optimal front or at least a reference set of the Pareto-optimal front. Usually, the convergence quality of a population is only measured by the non-dominated solutions which makes an evaluation according to the progress of the entire population impossible. Other metrics measure the convergence quality without referring to the set size. Therefore, a statistically reasonable indicator has been introduced which only requires the knowledge of an ideal point and is able to measure the convergence progress of the entire population.

This ACV indicator has several preferable properties that have been proved and discussed: ACV does not ignore the existence of multiple copies of the same solution within a solution set. It provides better indicator values to finer Pareto-optimal approximation sets and is compatible with the Pareto compliance as well as with the complete and strong outperformance relations. ACV is of a low computational complexity and simple to calculate. It has been used as convergence indicator as well as a discerning selection criterion. For these purposes, ACV has to be used in combination with a diversity measure to evaluate the performance of a MOEA in the first case or with a diversity preserving method in the case of a selection strategy. ACV is not able to focus on convergence and diversity simultaneously. A disadvantage of ACV as selection criterion is the guidance of the search process in the direction of one solution on the Pareto-optimal front that achieves the lowest cuboid volume to the ideal point.

Two selection strategies based on the ACV indicator have been proposed and are benchmarked on a generic three-dimensional biochemical minimization problem compared to the aggregate selection strategy, which uses a Pareto rank-based selection criterion. Further, the impact of the two selection parameters - the probability p_0 for the fitness-directional guidance and the tournament size - on the performance of the customized NSGA-II is determined.

The questions raised in the introduction are addressed in the following based on the experimental results: The first question refers to the identification of an optimal selection strategy for this optimization problem. In general, a good and comparable convergence–diversity balance is equally achieved by the three selection strategies with specific p_0 setting. The convergence–diversity performances of aggregate selection with $p_0 = 60\%$, of ACV-based selection with $p_0 = 50\%$, and of ACV-random selection with $p_0 = 10\%$ are comparable and present an ideal compromise between convergence and diversity. The main performance differences are observable on the relative non-dominated solution quality and are further discussed by the answer of the second question.

The second question refers to the influence of the parameter variation on the performance of the customized NSGA-II. An increase of the probability for the fitness-directional guidance results in an improvement of the convergence quality. At the same time, the diversity is decreased in the cases of aggregate selection and ACV-random selection. Therefore, a stringent fitness-directional guidance is achieved by the Pareto ranked-based principle in the case of aggregate selection and the ACV indicator based in the case of ACV-based and ACV-random selection. The main and important difference of the fitness-directional guidance strategies is observable by the relative non-dominated solution quality. The ACV_{rel} values of the selection strategies based on the ACV indicator are decreased which indicates an improvement of the non-dominated solution quality by comparable convergence results at the same time. In contrast, the ACV_{rel} results are increased by an increase of p_0. In general, the ACV_{rel} results of ACV-random selection are higher than those of ACV-based selection. The increase of the tournament size results in an improvement of the convergence quality in the case of aggregate selection and in an improvement of the diversity and relative non-dominated solution quality in the case of the ACV selection strategies.

The third question refers to generalization of the experimental results. In general, these selection strategies are generic and therefore transferable to other real-valued

MOEA. A similar performance of these selection strategies with the parameter variation is expected but has to be proved for other optimization problems which are part of the future work.

References

1. Okabe, T., Jin, Y., Sendhoff, B.: A critical survey of performance indices for multi-objective optimisation. In: Proceedings of the IEEE Congress on Evolutionary Computation, vol. 2, pp. 878885 (2003)
2. Grosan, C., Oltean, M., Dumitrescu, D.: Performance metrics for multiobjective optimization evolutionary algorithms. In: Proceedings of the Applied and Industrial Mathematics(2003)
3. Rosenthal, S., Borschbach, M.: Impact of population size and selection within a customized NSGA-II for biochemical optimization assessed on the basis of the average cuboid volume indicator. In: 6th International Conference on Bioinformatics, Computational Systems and Biotechnologies (BIOTECHNO 2014), IARIA, pp. 1–7 (2014)
4. Emmerich, M., Beume, N., Naujoks, B.: An EMO Algorithm Using the Hypervolume Measure as Selection Criterion. EMO 2005, LNCS 3410, 62–76 (2005)
5. Zitzler, E., Künzli, S.: Indicator-based selection in multiobjective search. Parallel Prob. Solving Nat. (PPSN VIII) **3242**, 832–842 (2004)
6. Rosenthal, S., El-Sourani, N., Borschbach, M.: Introduction of a Mutation Specific Fast Non-dominated Sorting GA Evolved for Biochemical Optimization. SEAL 2012, LNCS 7673, 158–167 (2012)
7. Rosenthal, S., El-Sourani, N., Borschbach, M.: Impact of Different Recombination Methods in a Mutation-Specific MOEA for a Biochemical Application. In: Vanneschi, L., Bush, W.S., Giacobini, M. (eds.) EvoBIO 2013, LNCS 7833, 188–199 (2013)
8. Rosenthal, S., Borschbach, M.: A benchmark on the interaction of basic variation operators in multi-objective peptide design evaluated by a three dimensional diversity metric and a minimized hypervolume. In: Emmerich, M. et. al. (eds.), EVOLVE - A Bridge between Probability, Set Oriented Numerics and Evolutionary Computation IV, pp. 139–153 (2013)
9. Hansen, M.P., Jaszkiewicz, A.: Evaluating the Quality of Approximations to the Non-dominated Set. Technical Report IMM-REP-1998-7, Technical University of Denmark (1998)
10. Knowles, J., Corne, D.: On metrics for comparing nondominated sets. In: Congress on Evolutionary Computation (CEC 2002), pp. 711–716. IEEE Press, New Jersey (2002)
11. Zitzler, E., Thiele L.: Multiobjective optimization using evolutionary algorithms—a comparative case study. In: Eiben, A.E., Bäck, T., Schoenauer, M., Schwefel, H.P. (eds.), Fifth International Conference on Parallel Problem Solving form Nature (PPSN-V), pp. 292–301. Berlin, Germany (1998)
12. Zitzler, E.: Evolutionary algorithms for multiobjective optimization: methods and applications. Ph.D. dissertation, Swiss Federal Institute of Technology (ETH) Zurich (1999)
13. Laumanns, L., Zitzler, E., Thiele, L.: A unified model for multi-objective evolutionary algorithms with elitism. In: CEC 2000, vol. 1, pp. 46–52 (2000)
14. Zitzler, E., Thiele, L., Laumanns, M., Fonseca, C.M., da Fonseca, V.G.: Performance assessment of multiobjective optimizers: An analysis and review. IEEE Trans. Evolut. Comput. **7**(2), 117–132 (2003)
15. Beume, N., Rudolph, G.: Faster S-metric calculation by considering dominated hypervolume as klees measure problem. Evol. Comput. **17**(4), 477–492 (2009)
16. Beume N., Rudolph, G.: Faster S-metric calculation by considering dominated hypervolume as Klee's measure problem. In: Proceedings of the Second IASTED Conference on Computational Intelligence, pp. 231–236 (2006)
17. Bradstreet, L., While, L., Barone, L.: A fast incremental hypervolume algorithm. IEEE Trans. Evolut. Comput. **12**(6), 714–723 (2008)

18. While, L., Bradstreet, L., Barone, L.A.: Fast way of calculating exact hypervolume. IEEE Trans. Evolut. Comput. **10**, 29–38 (2006)
19. Van Veldhuizen, D.A.: Multiobjective evolutionary algorithms: classification, analyses and new innovations. Ph.D. dissertation, Air Force Institute of Technology, Dayton, Ohio (1999)
20. Veldhuizen, D.A., Lamont, G.B.: Multiobjective evolutionary algorithm test. In: Carroll, J., Haddad, H., Oppenheim, D., Bryant, B., Lamont, G.B. (eds.) Proceedings of the 1999 ACM Symposium on Applied Computing, pp. 351–357. San Antonio, Texas (1999)
21. Deb, K., Jain, S.: Running Performance Metrics for Evolutionary Multiobjective Optimization. Kan GAL Report No. 2002004, Kanpur Genetic Algorithms Laboratory, Indian Institute of Technology Kanpur (2002)
22. Schütze, O., Esquivel, X., Lara, A., Coello Coello, C.A.: Using the averaged Hausdorff distance as a performance measure in evolutionary multiobjective optimization. IEEE Trans. Evolut. Comput. **16**(4), 504–522 (2012)
23. Coello Coello, C.A., Cruz Cortis, N.: Solving multiobjective optimization problems using an aritifical immune system. Genetic. Program. Evol. Mach. **6**(2), 163–190 (2005)
24. Trautmann, H., Wagner, T., Brockhoff, D.: Focused multiobjective search using R2-indicator-based selection. In: Learning and Intelligent Optimization, pp. 70–74. Springer, Berlin Heidelberg (2013)
25. Wagner, T., Trautmann, H., Brockhoff, D.: Reference articulation by means of the R2 indicator. In: Evolutionary Multi-criterion Optimization (EMO 2013), vol. 7811, 81–95 (2013)
26. Brockhoff, D., Wagner, T., Trautmann, H.: On the properties of the R2 indicator. In: Genetic and Evolutionary Computation Conference (GECCO 2012), pp. 465–472 (2012)
27. von der Lippe, M.: Deskriptive Statistik. Oldenburg Verlag (2006)
28. Zitzler, E., Brockhoff, D., Thiele, L.: The hypervolume indicator revisited: on the design of pareto compliant indicators via weighted integration. EMO **2007**, 862–876 (2007)
29. Deb, K., Pratap, A., Agarwal, S.: A fast and elitist multiobjective genetic algorithm: NSGA-II. IEEE Trans. Evolut. Comput. **6**(2), 182–197 (2002)
30. Rosenthal, S., Borschbach, M.: Impact of population size, selection and multi-parent recombination within a customized NSGA-II for biochemical optimization. Int. J. Adv. Life Sci. IARIA **6**(3&4), 310–324 (2014)
31. Bäck, T., Schütz, M.: Intelligent mutation rate control in canonical genetic algorithm. In: Proceedings of the International Symposium on Methodology for Intelligent systems, pp. 158–167 (1996)
32. BioJava: CookBook, release 3.0 http://www.biojava.org/wiki/BioJava. Cite January 2014
33. Cleland, J.L., Langer, R., Washington, D.C.: Formulation and Delivery of Proteins and Peptides: Design and Development Strategies, pp. 1–19. American Chemical Society (1994)
34. Hopp, T.P., Woods, K.R.: A computer program for predicting protein antigenic determinants. Mol. Immunol. **20**(4), 483–489 (1983)
35. Hansch, C., Björkroth, J., Leo, A.: Hydrophobicity and central nervous system agents: on the principle of minimal hydrohpobicity in drug design. J. Pharmacol. Sci. **76**(9), 663–687 (1987)
36. Rosenthal, S., Freisleben, B., Borschbach, M.: Aggregate selection in multi- objective biochemical optimization via the average cuboid volume indicator. In: Emmerich et al. (eds.) EVOLVE—A Bridge between Probability, Set Oriented Numerics and Evolutionary Computation VI (2015)

Printed in the United States
By Bookmasters